信息科学与技术丛书

产品级性能调优与故障诊断分析

郑 健 编著

机械工业出版社

本书根据作者多年的性能调优经验，以及客户实战案例归纳总结，形成了一套完整的性能优化方法，包括性能优化思路、代码效率分析方法、编码规范、服务器性能监控、客户实战案例、数据库性能分析及故障诊断方法、基于Web技术的性能优化方案等。

本书主要讲解产品级的性能调优技术，适合从事软件研发的开发人员、测试工程师（主要是白盒或集成并发测试人员）、DBA工程师、前线的技术支持工程师以及计算机系统维护人员。

另外，虽然本书是以.NET平台为案例展开讲解，但本质是讲解性能优化的分析思路和方法。不管在什么平台下，性能优化思想和方法都是相同的，只是一些具体的性能优化工具不同。

图书在版编目（CIP）数据

产品级性能调优与故障诊断分析/郑健编著.—北京：机械工业出版社，2015.2

（信息科学与技术丛书）

ISBN 978-7-111-49263-4

Ⅰ.①产… Ⅱ.①郑… Ⅲ.①计算机网络—程序设计 Ⅳ.①TP393.09

中国版本图书馆CIP数据核字（2015）第023337号

机械工业出版社（北京市百万庄大街22号 邮政编码100037）
策划编辑：周 萌　　责任校对：张艳霞
责任编辑：周 萌　　陶 韬
责任印制：李 洋

三河市宏达印刷有限公司印刷

2015年2月第1版·第1次印刷
184mm×260mm·11.25印张·276千字
0001—3000册
标准书号：ISBN 978-7-111-49263-4
定价：49.00元

凡购本书，如有缺页、倒页、脱页，由本社发行部调换

电话服务　　　　　　　　　　　网络服务
服务咨询热线：（010）88361066　机工官网：www.cmpbook.com
读者购书热线：（010）68326294　机工官博：weibo.com/cmp1952
　　　　　　　（010）88379203　教育服务网：www.cmpedu.com
封面无防伪标均为盗版　　　　　金 书 网：www.golden-book.com

前　言

最近几年我一直在做产品优化的工作，我打算写这本书的主要目的是把自己的知识沉淀记录下来，不仅在于对自己的总结，更在于分享。

一个运行中的系统会受到很多因素的制约，软件因素包括操作系统、数据库、服务器软件、通信协议、浏览器等，硬件因素包括 CPU、内存、磁盘、网络等。服务器硬件配置高，从某种程度上讲是高性能的前提，但性能优化要做的是用低功耗、低成本的技术运行高性能的产品，或者在当前服务器配置下让运行在其中的软硬件达到最佳的高性能状态，使软硬件资源消耗均衡，最终达到功能、体验、性能之间的完美平衡。

本书根据我多年的性能调优经验以及客户实战案例归纳总结，形成了一套完整的性能优化方法。工欲善其事，必先利其器，本书主要讲述对产品性能调优的各个方面，如服务器、客户端、数据库、Web 页面优化、服务器监控（CPU、内存、磁盘、网络等）、Web 服务器故障诊断，以及开发人员高性能编码规范等所有领域。通过阅读本书，可以让您全面掌握主流 C/S 及 B/S 架构下产品级调优方案。

一个产品开发完成并不意味着软件生命周期的结束，还要考虑对客户产品的后期维护成本。很多产品在开发阶段用的时间并不多，但维护阶段出现异常或性能问题时可能会消耗几倍于开发周期的时间。除了优化技术外，本书也可以让您掌握一套服务器疑难问题诊断方法，这样遇到棘手问题时就有思路去分析定位并解决。

目前市场上讲解开发的书籍很多，但讲解软件产品级性能调优的书籍比较少，即使有一些讲解调优的书籍可能也只讲到某一个领域，比如在 SQL Server 领域调优。事实上在客户现场，导致性能问题的因素有很多，比如客户端脚本、服务端代码、网络环境、服务器故障、数据库瓶颈，还有问题最多的服务器，尤其是 Web 服务器运行中崩溃/异常、内存泄露、CPU 占用率高、线程死锁挂起等不可预测的故障。换句话说，客户在使用软件时遇到问题后会跟我们说"卡死了""运行慢了""没反应了""CPU 100%了"，而不会跟我们说哪个部件出问题了，比如"数据库慢了""客户端 Java Scirpt 脚本慢了""遇到网速瓶颈了""磁盘有队列了""数据库占用 100% CPU 了"，所以这就要求我们对所有可能的领域非常熟悉，并且有丰富的经验，能够根据现象推测是哪个软/硬件出了问题，然后再对相应部件进行进一步定位。而本书中就是对上面提到的这些领域进行性能分析及疑难故障诊断。

本书主要讲解产品级的性能调优技术，适合软件研发人员；测试工程师（不是黑盒测试人员，主要是对白盒或集成并发测试人员）；从事数据库维护的 DBA 工程师；客户前线的技术支持工程师；计算机系统维护人员等。另外，虽然本书是以.NET 平台为案例展开讲解，本质是讲解性能优化的分析思路和方法，任何平台下都可以融汇贯通。

编者

目 录

前言
第1章 性能优化思路 ··· 1
  1.1 两个优化实战案例 ·· 2
    1.1.1 内存性能问题案例 ··· 2
    1.1.2 CPU 占用 100%分析案例 ··· 6
  1.2 性能优化理论体系 ·· 12
第2章 代码效率分析方法 ·· 15
  2.1 服务端代码性能分析方法 ·· 16
    2.1.1 VSTS 性能分析工具 Profiler 介绍 ·· 16
    2.1.2 VSTS 性能分析工具使用 ·· 16
    2.1.3 VSTS 报表字段字典 ·· 22
  2.2 客户端代码性能分析方法 ·· 26
    2.2.1 IE8 Profiler 简介 ·· 26
    2.2.2 使用 IE8 Profiler 分析客户端脚本 ·· 26
    2.2.3 查看 IE8 Profiler 分析报告 ·· 27
    2.2.4 IE8 Profiler 报表字段字典 ··· 29
  2.3 性能调优工具集锦 ·· 30
第3章 编码规范 ··· 32
  3.1 概述 ·· 33
  3.2 编码规范 ·· 33
    3.2.1 数据库设计及编码规范 ·· 33
    3.2.2 客户端代码编码规范 ··· 39
    3.2.3 服务器代码编码规范 ··· 44
第4章 服务器性能监控 ··· 50
  4.1 概述 ·· 51
  4.2 服务器性能监控 ··· 52
    4.2.1 内存 ··· 52
    4.2.2 处理器 ·· 53
    4.2.3 磁盘 ··· 54
    4.2.4 网络 ··· 54
    4.2.5 进程 ··· 56
    4.2.6 系统 ··· 56
    4.2.7 .NET CLR Memory ··· 57
    4.2.8 .NET CLR Loading ··· 57
    4.2.9 Asp.net ·· 57

| | 4.2.10 | 数据库 ··· | 57 |

第 5 章　客户实战案例 ··· 60
- 5.1　概述 ··· 61
- 5.2　WinDbg 工具介绍 ··· 61
  - 5.2.1　环境配置 ··· 61
  - 5.2.2　常用命令简介 ·· 62
  - 5.2.3　示例应用 ··· 65
- 5.3　客户问题诊断案例 ··· 70
  - 5.3.1　Web 服务器内存达到 3GB 后崩溃原因诊断定位 ·· 70
  - 5.3.2　Web 服务器运行中突然崩溃原因定位 ··· 78
  - 5.3.3　DevGrid 控件 EventHandler 事件泄漏内存 ··· 83
  - 5.3.4　Session 陷阱及正确使用 ·· 86
  - 5.3.5　WinDbg 内存泄漏+异常检测案例 ··· 89

第 6 章　数据库性能分析及故障诊断方法 ·· 103
- 6.1　数据库优化概述 ··· 104
- 6.2　效率专题研究 ··· 104
  - 6.2.1　数据库无法收缩变小原因分析案例汇总 ·· 104
  - 6.2.2　数据库碎片增长过快原因分析及建议方案 ·· 110
  - 6.2.3　聚集索引对插入效率的影响 ··· 113
  - 6.2.4　多表连接方案效率评估 ·· 114
- 6.3　优化方法指令 ··· 116
  - 6.3.1　显示查询计划 ·· 116
  - 6.3.2　查看 SQL 内部执行计划生成/优化信息 ·· 117
  - 6.3.3　查看缓存对象（syscachobjects） ·· 117
  - 6.3.4　清空缓存 ··· 117
  - 6.3.5　STATISTICS IO ··· 118
  - 6.3.6　STATISTICS TIME ·· 118
  - 6.3.7　分析执行计划 ·· 118
  - 6.3.8　索引优化 ··· 121
  - 6.3.9　数据库和文件空间 ··· 128
  - 6.3.10　监视命令 ··· 132
  - 6.3.11　SQL 性能统计 ··· 133
  - 6.3.12　跟踪文件统计 ·· 134
- 6.4　性能故障检测方法 ··· 136
  - 6.4.1　CPU 问题诊断 ··· 136
  - 6.4.2　内存诊断 ··· 138
  - 6.4.3　I/O 诊断 ·· 141
  - 6.4.4　tempdb 诊断 ··· 143
  - 6.4.5　阻塞诊断 ··· 144

  6.4.6 死锁诊断 ························ 147
  6.4.7 排除故障 ························ 154
  6.4.8 信息查询 ························ 155
  6.4.9 存储引擎 ························ 155

第 7 章 基于 Web 技术的性能优化方案 ·················· 165
7.1 Web 技术优化方案 ························ 166
  7.1.1 发布时要关闭调试模式 ···················· 166
  7.1.2 服务器和客户缓存利用 ···················· 166
  7.1.3 启用 GZIP 压缩功能 ····················· 166
  7.1.4 对站点中的静态资源精简与压缩 ················· 166
  7.1.5 JavaScript/CSS 输出位置规范 ·················· 167
  7.1.6 减少页面请求 ······················· 168
  7.1.7 禁用服务器控件的视图状态 ··················· 169
  7.1.8 定制仅满足特定功能的自定义控件 ················ 169
  7.1.9 优化方案提升数据 ····················· 169
7.2 网络瓶颈诊断 ·························· 169
  7.2.1 各种网速测试方法 ····················· 169
  7.2.2 网络瓶颈诊断 ······················· 171

第 1 章
性能优化思路

本章内容
- 内存性能优化案例
- CPU 占用 100%分析案例
- 性能优化理论体系

1.1 两个优化实战案例

在本书的第一章，我并不喜欢先说一些理论上的东西，这样会让读者感到乏味，本书先以两个真实的性能问题案例开篇，让读者了解一下定位一个性能问题的过程，或许这样会更有趣。

1.1.1 内存性能问题案例

1．客户问题描述

客户反馈查询报表速度太慢，要十几分钟，而且是所有报表查询都慢，其他一些轻量级操作正常。客户服务器配置如表 1-1 所示。

表 1-1　客户服务器配置

硬件类型	参数值
CPU	至强 8 核
内存	32GB
磁盘	SAS（>10000RPM）
操作系统	Microsoft windows server 2003 sp2, 32 位
数据库	Microsoft SQL Server 2005 sp2, 32 位

2．诊断分析定位原因

根据客户的反馈及服务器环境，提取出以下三个线索：
- 只有报表查询慢，而报表查询对各硬件资源消耗比较大。
- 客户服务器当前环境安装的是 32 位的操作系统及 32 位的数据库版本。
- 客户服务器配置了 32GB 的内存。

首先会想到的是内存问题，先从内存入手解决。

用 Windows Performance 计数器对运行中的服务器做了日志跟踪，拿到本地并打开，如图 1-1 所示。

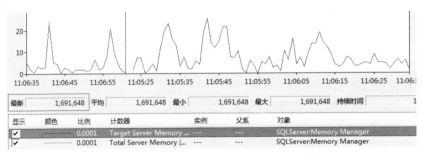

图 1-1　服务器日志跟踪

数据库目标内存（Target Server Memory）只有 1.6GB，指的是操作系统当前分配给数据库可用的最大内存（即使设置了 32GB，如果有内存不够或权限问题等原因，数据库也不会用到 32GB 内存，后面会说明）。当前总共使用内存（Total Server Memory）指的是当前数据库已经使用了多少内存，本例监控到的值也是 1.6GB，也就是说已经用完了当前所有目标内存。

另外，数据库设置的最大内存是 30GB（并打开 AWE 功能），如图 1-2 所示。

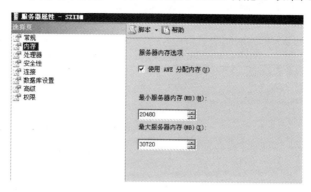

图 1-2　内存设置

客户给 SQL Server 设定的最大使用内存是 30GB，并且开启了 AWE 功能，但为什么内存实际只用了 1GB 呢？

这是因为：虽然客户设置了 30GB，并且设置了 AWE，但这里的 AWE 并没有生效。

操作系统和数据库都安装的是 32 位的版本，尽管配置了 32GB 内存，但是 32 位操作系统最多只能识别 4GB 的内存，所以内存使用受到限制。在这 4GB 内存中，默认情况下操作系统会留 2GB 给自己（内核）用，剩下的 2GB 内存会给其他所有应用程序用，也就是说 SQL Server 不可能使用超过 2GB 的内存，跟之前客户服务器上内存使用的 1.6GB 相吻合。即使打开 3GB 开关，也仅有 3GB 给所有应用程序使用，32GB 物理内存也浪费了。

最佳解决方案：

以下经验仅针对数据库服务器内存配置建议：

- 内存大于 4GB

对于内存超过 4GB 的服务器，建议安装 64 位的操作系统和 64 位的数据库版本，这样可以避免 32 位版本的 4GB 内存限制，也不需要做本节中的内存设置优化（PAE/AWE/内存锁定页权限）。

这个建议仅在开始部署环境时用得多，对当前已经上线的系统，则用得最多的还是进行内存设置优化（PAE/AWE/内存锁定页权限）。

- 内存小于或等于 4GB

对于内存等于或小于 4GB 的物理内存配置，建议安装 32 位的操作系统和数据库版本并进行内存设置优化（PAE/AWE/内存锁定页权限）。64 位的操作系统下所有程序比较耗内存，4GB 物理内存有点小。比如数据库服务器在 32 位环境下并配置 4GB 物理内存，如果未开启 PAE 和 AWE，数据库最大会分配到 1.6GB 左右内存；如果开启了 PAE 和 AWE，则数据库使用内存大概在 2.8GB 内存，64 位操作系统下不会达到 2.8GB。

继续回到客户问题。由于当前系统已经上线，并且客户正在使用，重装系统和数据库环境可能不太现实，最佳方案是进行内存设置优化（PAE/AWE/内存锁定页权限）。具体步骤如下：

第一步，开启操作系统 PAE（物理扩展内存），配置系统盘下的 boot.int 文件。

```
[boot loader]
timeout=30
default=multi(0)disk(0)rdisk(0)partition(2)\WINDOWS
[operating systems]
multi(0)disk(0)rdisk(0)partition(2)\WINDOWS="Windows Server 2003, Enterprise" /fastdetect /PAE
```

直接在 boot.ini 文件中增加 /PAE 参数。

如果是 Windows7，Windows Server 2008 及更高版本的操作系统，操作系统提供了专门的 BCDEdit 命令：

```
BCDEdit /set PAE forceenable
```

第二步，开启数据库的 AWE（Address Windowing Extensions）动态分配映射内存，勾选"使用 AWE 分配内存"，并设置最大服务器内存为 30720（30GB），如图 1-3 所示。

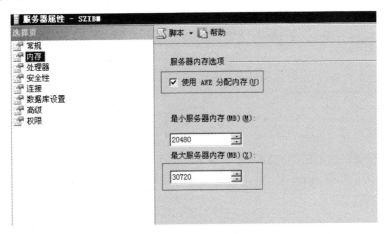

图 1-3　设置最大内存

第三步，设置内存锁定页权限。首先确定一下 SQL Server 当前运行账户，如图 1-4 所示。

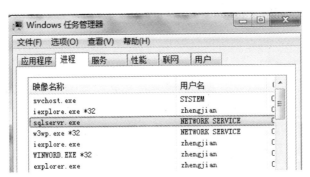

图 1-4　任务管理器

当前运行账户为：NETWORK SERVICE 网络用户，后面步骤就为此用户赋予内存锁定页权限。

运行 gpedit.msc 命令，如图 1-5 所示，打开"本地组策略编辑器"。

图 1-5 运行窗口

在"本地组策略编辑器"中依次展开左侧目录结点,找到"用户权限分配"结点,如图 1-6 所示。

图 1-6 设置锁定内存页

右击"锁定内存页",单击"属性"选项,打开"锁定内存页 属性"窗口,如图 1-7 所示。

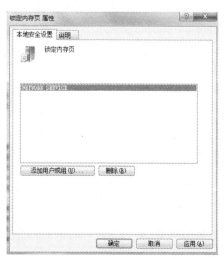

图 1-7 添加锁定内存页权限账户

5

把 SQL Server 当前登录用户 NETWORK SERVICE 添加进来，单击"确定"按钮。

注意：这三个步骤缺一不可，设置好后要重启一下计算机。

经过此设置后，几分钟后再看客户服务器的计数器，如图 1-8 所示。

图 1-8　查看最大内存

目标内存已经由 1.6GB 变为 30GB，当前使用内存也由原来的最大使用 1.6GB 变为 3.8GB，接下来这个值还会增加，上限是目标内存最大值。

客户报表查询速度也非常快了，整体上操作都响应非常快了。问题原因主要在于客户配置了 32GB 物理内存，但数据库只使用了 1.6GB。

问题思路：

第一步，先确定环境问题（操作系统、数据库配置等）。

第二步，如果环境没问题，再进行报表服务端代码或报表查询 SQL 跟踪。

不要直接进入第二步，这样分析方向就错了。一个性能专业人员，除了要掌握专业的解决问题技能外，积累经验也非常重要。

下面是一个更有趣的案例。

1.1.2　CPU 占用 100%分析案例

1．客户问题描述

200 人并发使用某服务器，使用中出现所有客户端卡死，服务器无法接收客户端任何请求。

客户还提供了一个线索：此时服务器 CPU 利用率接近 100%。

2．定位分析

这台服务器运行着 ERP 系统，主要承载 Web 服务器，根据客户提供的线索很可能是 CPU 利用率 100%导致服务器繁忙，而不能及时响应所有客户端的请求，出现所有客户端"假死"现象，这是很常见的问题。

往往很多非专业计算机人员遇到这种问题就重启一下 Web 服务器或客户服务器，继续使用。这样既解决不了问题，又会丢失线索，而且问题还会重复出现。

先看一下服务器计数器，计数器中显示"% Time in GC"为 CPU 利用率 90%，如图 1-9 所示。

图 1-9　垃圾回收线程 CPU 利用率

说明：一般 % Time in GC > 10%，基本上就应该检查代码了，而这里达到 90%。

现象已经基本明确，是由于 w3wp 中的 GC 线程不断地在做垃圾回收工作，耗尽 CPU 资源，导致服务器不能处理其他客户端客户发来的请求。

到这里只是笼统的分析，还不能确定是什么问题，更不能做任何结论，要确定是什么导致 GC 这么忙碌才是最终目的。一般涉及 GC 问题都是服务端代码写法不正确导致的，找到代码并修改才是根本。下面就分析一下是哪句代码出的问题。

这里使用 WinDbg 工具从 webserver 的进程中寻找线索。WinDbg 是微软内部用来调试操作系统 bug 的一个工具，当然也能够调试应用程序软件。如果读者对这个工具不熟悉也没有关系，这里只是说明一下分析思路，这一章中还不需要对每个分析点具体了解。

接到客户问题后，使用 WinDbg 对服务器进程 w3wp.exe dump 了一个文件，把 dump 文件拿回本地分析，重启一下服务器，客户可继续使用系统。

Dump 文件的过程是把应用运行中某一时刻的运行信息及状态写到文件中，查看一下线程池：

```
0:025> !threadpool
CPU utilization: 99%
Worker Thread: Total: 47 Running: 5 Idle: 42 MaxLimit: 800 MinLimit: 8
Work Request in Queue: 0
--------------------------------------
Number of Timers: 46
--------------------------------------
Completion Port Thread:Total: 2 Free: 2 MaxFree: 16 CurrentLimit: 2 MaxLimit: 800 MinLimit: 8
```

说明：在本书中由于这样的分析展示比较多，关键数字我会以粗体进行标注，比如"**99%**"被标记为粗体显示。

CPU 利用率比较高，dump 文件这一刻 CPU 利用率是 99%，说明很可能在这个 dump 文件中能够找到线索。

w3wp 通过内部多线程方式来同时处理多客户端请求，线程池的数量根据请求数自动分配，一般有几十个线程在工作。随便选择一个线程：

```
0:025> kb
ChildEBP RetAddr  Args to Child
1b24f974 75430816 000003c0 00000000 00000000 ntdll!ZwWaitForSingleObject+0x15
1b24f9e0 76da1194 000003c0 ffffffff 00000000 KERNELBASE!WaitForSingleObjectEx+0x98
1b24f9f8 6c3f1030 000003c0 ffffffff 00000000 kernel32!WaitForSingleObjectExImplementation+0x75
1b24fa2c 6c3f1071 000003c0 ffffffff 00000000 clr!CLREvent::CreateManualEvent+0xf6
1b24fa7c 6c3ed3e8 00000000 75393cd7 00000000 clr!CLREvent::CreateManualEvent+0x137
1b24fabc 6c3ed409 ffffffff 00000000 00000000 clr!CLREvent::WaitEx+0x126
1b24fad0 6c4391dd ffffffff 00000000 00000000 clr!CLREvent::Wait+0x19
1b24faf0 6c43a296 1afa0048 00000002 6c43a370 clr!SVR::t_join::join+0xef
1b24fb10 6c43a08f 00000002 1b24fb30 00000001 clr!SVR::gc_heap::scan_dependent_handles+0x31
1b24fb58 6c439615 00000002 00000000 1afa057c clr!SVR::gc_heap::mark_phase+0x427
1b24fb84 6c439cbb 75393dab 00000004 1afa0048 clr!SVR::gc_heap::gc1+0x63
```

```
1b24fbc0 6c439328 00000000 00000000 1afa0048 clr!SVR::gc_heap::garbage_collect+0x30d
1b24fbe8 6c4998cb ffffffff 7754a11c 7754a0ca clr!SVR::gc_heap::gc_thread_function+0x73
1b24ff00 76da33ca 1afa0048 1b24ff4c 77549ed2 clr!SVR::gc_heap::gc_thread_stub+0x7e
1b24ff0c 77549ed2 1afa0048 6c4152e5 00000000 kernel32!BaseThreadInitThunk+0xe
1b24ff4c 77549ea5 6c499879 1afa0048 ffffffff ntdll!__RtlUserThreadStart+0x70
1b24ff64 00000000 6c499879 1afa0048 00000000 ntdll!_RtlUserThreadStart+0x1b
```

这个线程的调用堆栈如上所示，调用顺序从下往上。可以看到这个线程在等待 GC 操作。又看了其他几个不同的线程，也是如此。

其中有个 35 号线程有点问题，它的堆栈调用如下：

```
0:025> ~35s
eax=00000000 ebx=00000000 ecx=27591b10 edx=1576c8c0 esi=000003f8 edi=00000000
eip=7752f8c1 esp=1c67d518 ebp=1c67d584 iopl=0         nv up ei pl zr na pe nc
cs=0023  ss=002b  ds=002b  es=002b  fs=0053  gs=002b              efl=00000246
ntdll!ZwWaitForSingleObject+0x15:
7752f8c1 83c404          add     esp,4
0:035> !clrstack
OS Thread Id: 0x17e8 (35)
Child SP IP     Call Site
1c67d820 7752f8c1 [HelperMethodFrame: 1c67d820]
1c67d870 6b87781c System.String.Concat(System.String, System.String)
1c67d888 0100cb90 U.King.EE.FF.EEDoc.DocFF.GetAllSubEmployee(System.String)
1c67d8b8 0100ca6d U.King.EE.Service.EEDoc.EEService.GetAllSubEmployee(System.String)
1c67de14 6c3921db [DebuggerU2MCatchHandlerFrame: 1c67de14]
1c67e08c 6b87d37c System.Reflection.RuntimeMethodInfo.Invoke(System.Object, System.Reflection
BindingFlags, System.Reflection.Binder, System.Object[], System.Globalization.CultureInfo, Boolean)
……（省略完整代码）
69c52cdd System.Web.HttpApplication.System.Web.IHttpAsyncHandler.BeginProcessRequest(System.
Web.HttpContext, System.AsyncCallback, System.Object)
1c67f0ac 69c9a8f2 System.Web.HttpRuntime.ProcessRequestInternal(System.Web.HttpWorkerRequest)
1c67f0e0 69c9a63d System.Web.HttpRuntime.ProcessRequestNoDemand(System.Web.HttpWorker
Request)
1c67f0f0 69c99c3d System.Web.Hosting.ISAPIRuntime.ProcessRequest(IntPtr, Int32)
1c67f0f4 6a2b5a7c [InlinedCallFrame: 1c67f0f4]
1c67f168 6a2b5a7c DomainNeutralILStubClass.IL_STUB_COMtoCLR(Int32, Int32, IntPtr)
1c67f2fc 6c3925c1 [GCFrame: 1c67f2fc]
1c67f36c 6c3925c1 [ContextTransitionFrame: 1c67f36c]
1c67f3a0 6c3925c1 [GCFrame: 1c67f3a0]
1c67f4f8 6c3925c1 [ComMethodFrame: 1c67f4f8]
```

一般%Time in GC 消耗时间比较大的原因是，方法长时间执行，并且产生很多对象，导致 GC 线程不断地释放空引用对象。而这里的 GetAllSubEmployee 方法内部存在连接运算 String.Concat（字符"+"），可以推测很可能是 GetAllSubEmployee 方法中有大量循环调用"+"的操作，导致不断地创建对象，不断地被 GC 线程回收，所以 GC 线程忙碌。到目前这只是推测。

进一步看一下 35 号线程中方法的参数值，如下：

```
0:035> !clrstack -a
OS Thread Id: 0x17e8 (35)
Child SP IP        Call Site
1c67d820 7752f8c1 [HelperMethodFrame: 1c67d820]
1c67d870 6b87781c System.String.Concat(System.String,    System.String)
    PARAMETERS:
        str0 (<CLR reg>) = 0x29c20038
        str1 (<CLR reg>) = 0x0f5c59f4
    LOCALS:
        0x1c67d870 = 0x003c2e10
        <no data>

1c67d888 0100cb90 U.King.EE.FF.GLDoc.DocFF.GetAllSubEmployee(System.String)
    PARAMETERS:
        this = <no data>
        strWhere = <no data>
    LOCALS:
        <no data>
        <no data>
        <no data>
        0x1c67d88c = 0x29c20038
        <no data>
        0x1c67d888 = 0x0f20fa84
        <no data>

1c67d8b8 0100ca6d U.King.EE.Service.GLDoc.DocService.GetAllSubEmployee(System.String)
    PARAMETERS:
        this = <no data>
        strWhere = <no data>
……（省略完整代码）
```

可以看到，String.Concat 的两个参数的地址分别为 0x29c20038 和 0x0f5c59f4，它们是 CLR 寄存器中存储数据的内存地址，通过这两个地址我们能够知道存储的是什么数据。

先看一下 0x0f5c59f4 指针中的数据：

```
0:035> !do 0x0f5c59f4
Name:         System.String
MethodTable: 6b8df9ac
EEClass:      6b618bb0
Size:         92(0x5c) bytes
File:         C:\Windows\Microsoft.Net\assembly\GAC_32\mscorlib\v4.0_4.0.0.0__b77a5c561934e089\mscorlib.dll
String:       ,'237fb7ed-99d8-4b8c-8ee5-49d4b848dc3d'
Fields:
      MT    Field   Offset              Type VT     Attr    Value Name
6b8e2978  40000ed        4       System.Int32  1 instance      39 m_stringLength
6b8e1dc8  40000ee        8       System.Char   1 instance      2c m_firstChar
6b8df9ac  40000ef        8      System.String  0   shared      static Empty
```

>> Domain:Value   01132680:0aaa0260 1afe6ae8:0aaa0260 <<

这个地址中存储的数据为粗体标注的 String 节部分，值为一个 guid 类型的数据。
再看一下 0x29c20038 指针存储的数据：

0:035> !do 0x29c20038
Name: System.String
MethodTable: 6b8df9ac
EEClass: 6b618bb0
Size: 7887918(0x785c2e) bytes
File: C:\Windows\Microsoft.Net\assembly\GAC_32\mscorlib\v4.0_4.0.0.0__b77a5c561934e089\mscorlib.dll
String: <String is invalid or too large to print>

Fields:
MT	Field	Offset	Type	VT	Attr	Value	Name
6b8e2978	40000ed	4	System.Int32	1	instance	**3943952**	m_stringLength
6b8e1dc8	40000ee	8	System.Char	1	instance	27	m_firstChar
6b8df9ac	40000ef	8	System.String	0	shared	static	Empty

>> Domain:Value   01132680:0aaa0260 1afe6ae8:0aaa0260 <<

值为 "<String is invalid or too large to print>"，值已经太大了，不能显示，估算了一下字符串长度为近 4MB（长度：3943952）。

还是得看一下这个地址 0x29c20038 的值才行，性能优化分析不能凭猜，根据我的经验一般去猜十有九错，要用数据说话。用另一个内存查看命令显示一下它的值：

0:035> du 0x29c20038 0x29c20038+1000
29c20038   "18c35a15-6a48-40fd-b4ad-001"
29c20078   "7ddafa85d','18c35a15-6a48-40fd-b"
29c200b8   "4ad-0017ddafa85d','18c35a15-6a48"
29c200f8   "-40fd-b4ad-0017ddafa85d','18c35a"
29c206b8   "-6a48-40fd-b4ad-0017ddafa85d','1"
29c206f8   "8c35a15-6a48-40fd-b4ad-0017ddafa"
29c20738   "85d','18c35a15-6a48-40fd-b4ad-00"
29c20a38   "99012f2b6','cda9b452-8ec0-416c-a"
29c20a78   "436-00299012f2b6','cda9b452-8ec0"
29c20ab8   "-416c-a436-00299012f2b6','cda9b4"
29c20af8   "52-8ec0-416c-a436-00299012f2b6',"
29c20b38   "'cda9b452-8ec0-416c-a436-0029901"
29c20b78   "2f2b6','cda9b452-8ec0-416c-a436-"
29c20bb8   "00299012f2b6','cda9b452-8ec0-416"
29c20bf8   "c-a436-00299012f2b6','cda9b452-8"
29c20c38   "ec0-416c-a436-00299012f2b6','cda"
29c20c78   "9b452-8ec0-416c-a436-00299012f2b"
29c20cb8   "6','cda9b452-8ec0-416c-a436-0029"
29c20cf8   "9012f2b6','cda9b452-8ec0-416c-a4"
29c20d38   "36-00299012f2b6','cda9b452-8ec0-"

29c20d78	"416c-a436-00299012f2b6','cda9b45"
29c20db8	"2-8ec0-416c-a436-00299012f2b6','"
29c20df8	"cda9b452-8ec0-416c-a436-00299012"
29c20e38	"f2b6','cda9b452-8ec0-416c-a436-0"
29c20e78	"0299012f2b6','cda9b452-8ec0-416c"
29c20eb8	"-a436-00299012f2b6','cda9b452-8e"
29c20ef8	"c0-416c-a436-00299012f2b6','cda9"
29c20f38	"b452-8ec0-416c-a436-00299012f2b6"
29c20f78	"','cda9b452-8ec0-416c-a436-00299"
29c20fb8	"012f2b6','cda9b452-8ec0-416c-a43"
29c20ff8	"6-00299012f2b6','cda9b452_8ec0-4"

……（省略完整代码）

……（省略完整代码）

表格中显示的即为指针 0x29c20038 的值，跟猜测完全一致。这里 4MB 长度的字符串不会有多大问题，问题是这里会产生成千上万个 guid 对象（**3943952/36=产生 109000 多个对象，也相当于这个 for 循环了 10.9 万次以上**），让 w3wp 的 GC 线程一直忙碌，不能处理客户端请求，并且消耗了 99%的 CPU 资源。

到现在问题原因已经非常明确，35 号线程 GetAllSubEmployee 方法中代码的原因导致此问题。接下来的工作就好做了，根据 GetAllSubEmployee 这个线索用 Reflector 反射代码看一下具体是怎么写的：

```
public string GetAllSubEmployee(string strWhere)
{
    string strSQL = this.GetWhereSql(strWhere);
    DataTable table = execute.query(strSQL);
    string strResult = "";
    foreach (DataRow row in table.Rows)
    {
        strResult = strResult + ((strResult != "") ? ("," + row["ID"].ToString() + "'") : ("'" + row["ID_EE"].ToString() + "'"));
    }
    if (strResult == "")
    {
        strResult = "'00000000-0000-0000-0000-000000000000'";
    }
    return strResult;
}
```

上面 foreach 中果真是采用 "+" 号进行连接的。循环了 10.9 万次以上，并且每次循环还用了多个字符串 "+" 连接符，而且 dump 文件时它还在循环过程中，可能实际循环次数还要多。仅仅这一句代码就导致 GC 线程占用 90%的 CPU，并导致所有客户端卡死及 CPU 占用率 100%。

另外，根据这个循环次数可以推断，可能是开发人员把数据库表中的所有数据追加起来，仅在客户使用过程中数据量增大后才呈现，该问题非常隐蔽。

还有，用 guid 数据库类型表示这样的业务，就不如用 int 或 string，guid 更浪费空间。

修改方案：采用 StringBuilder 代替 String.Concat。这再简单不过了，只要是开发人员都会知道怎么修改。

> 小提示：
> 为什么 StringBuilder 代替 String.Concat 会解决此问题呢？
> 这与 StringBuilder 的实现原理有关，它的类定义如下（伪代码）：
>
> Public class StringBuilder
> {
>   String strCurrentString;
>   StringBuilder orinal;
> }
>
> 当调用 StringBuilder 的 Append 方法时，StringBuilder 对象不会复制新的对象，而是生成新的 StringBuilder 对象，并用里面的 orinal 指向原来的对象。相当于用引用（指针）指向所有零散字符串的地址。简单地说，每次有新对象追加，只是改变一下指针，指向原来的对象地址。
> 而 String.Concat 则不一样，c=a+b 中的 "=" 运算符每运算一次总是会生成新的对象；另外 "=" 运算符也比较耗时，不如 StringBuilder.Append 方法效率高。

上面是对成功解决一个性能问题后的过程回放，事实上在找到问题之前分析的过程就像大海捞针，甚至毫无头绪，因为产生同一种现象的原因太复杂了。

这是一个根据现场痕迹寻找线索，再逆向推断定位，最终找到问题代码的过程，同时也是一个很有趣、很具有挑战性的过程。

1.2 性能优化理论体系

性能优化的理论体系如图 1-10 所示。

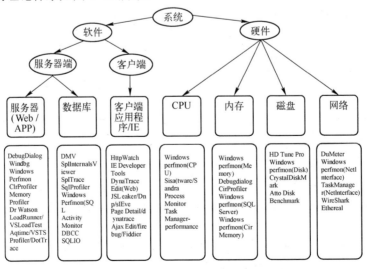

图 1-10 性能优化理论体系

说明：图1-10所示的这些工具集的应用场景在第2章还会有介绍。

当一个系统出现性能问题，概括地讲，一般主要由两种类型原因导致：系统上运行的软件和硬件。

软件部分又分为：Web服务器、应用服务器、数据库、客户端应用程序或浏览器等；硬件部分一般出现问题较多的有CPU、内存、磁盘、网络这4个硬件。每一个部件出现的问题也是各种各样的：

- Web服务器出现的问题比较典型，如：运行中崩溃/异常、内存泄露、CPU High、线程死锁挂起及常用的服务端代码性能问题等。
- 数据库服务器问题也较多，SQL脚本效率低下及缺少索引都可能会导致CPU、内存、磁盘、网络出现问题。

不管是Web服务器还是数据库服务器，它们都会导致CPU、内存、磁盘、网络等出现问题，并且出现问题的现象是一样的。可以说性能优化工程师就是一个医生，只是他诊断的不是病人，而是计算机系统；或者说是一个特工，在服务器崩溃现场寻找线索。

不管是软件还是硬件出现问题，作为一个做优化的专业人员，都应当有一些手段和工具去寻找、定位。

事实上在分析过程中，并不是如图1-10中标注的每个组件出现问题就用下方对应的手段（工具）去解决，也会遇到以下几种情况：

- 软件和硬件是互相影响的

软件有问题会导致硬件出现问题，硬件有问题会导致软件性能出现问题。比如我们在做产品监控时，监控到内存出现问题后，相应的CPU、磁盘等也会出现问题，要通过经验正确地找到哪个硬件是主要问题。

不管是软件还是硬件出现问题，都表现在硬件负载异常。比如磁盘出现问题了，有可能是缺少索引导致大量I/O所致，有可能是SQL效率低下所致，也有可能是数据库目标内存设置得太小导致内存与磁盘之间大量页交换所致……。一般来讲，大部分性能问题要先从软件入手，大部分问题可以通过软件部件优化解决。如果仍无法解决，再通过一些工具检测一下出现负载异常的硬件是否真的有问题，如CPU是否运算能力比较差，磁盘是否读写能力差，升级硬件解决即可。

- 杠杆平衡原理（内存⇔CPU和磁盘）

杠杆平衡原理主要是针对图1-10中右边的几个硬件而言。在一个系统中，内存配置越大，则系统对CPU和磁盘资源占用就越少；反之，内存配置越小，则系统对CPU和磁盘资源占用就越多。以数据库服务器为例，内存配置小了，系统内存不够用，就无法利用缓存，每次取数据都要从磁盘读取，并不断用置换算法把内存空间腾出来，缓存新的数据，就会不断导致大量的页交换，进而导致对CPU和磁盘更多的占用。

了解这个原理不管对我们做产品优化还是对客户资金投入都是有参考意义的。内存价格非常便宜，而CPU和磁盘都比内存贵得多，所以能够根据此经验给客户提供最佳性价比的服务器配置建议非常重要。

最后再补充一点，性能优化工作不仅仅是解决代码和SQL问题，要掌握一整套性能分析方法，包括对各种硬件和系统运行的环境都要熟悉。

例如数据库方面，开发工程师只要了解 SQL 语法基本上就可以开发了，但优化工程师除了要掌握这些，还要了解数据库的各个层面，如：存储引擎、编译优化、大并发下的阻塞/死锁诊断，及 CPU/内存/磁盘负载下的各种诊断方法。

在分析问题时最重要的不是工具应用得多么熟练，而是思考。本章的两个案例分别从不同的角度入手，最终解决问题，分析时不要搞错方向。

性能优化 = 思考 + 经验 + 方法

第 2 章
代码效率分析方法

本章内容
- 服务端代码性能分析方法
- 客户端代码性能分析方法
- 性能调优工具集锦
- 总结

2.1 服务端代码性能分析方法

服务端性能分析工具比较常用的有 VSTS Profiler、AQTime、DotTrace 等，本节主要以 VSTS Profiler 为例。

这章的内容是性能优化技术，也是对开发人员最基本的技能要求。

2.1.1 VSTS 性能分析工具 Profiler 介绍

VSTS Profiler 可以分析任何服务端代码执行时间，当然也包括服务器控件服务端代码。当产生性能瓶颈时，即可以使用此工具检测是服务端哪个方法导致了性能瓶颈。

使用 VSTS Profiler，最低要求必须安装：Visual Studio Team System 2008 + Microsoft .NET Framework V3.5 SP1，或 VS2010 等更高版本。

VSTS 里面集成的工具 Profiler，可以帮助研发人员在程序运行的过程中收集相关的数据，并且对其进行分析，从而达到帮助实现性能调优的目的。

在 VSTS Profiler 中，有两种 Profiling 的方法：

第一种是采样（Sampling）。采样模式的工作原理是 Profiler 定期中断 CPU，并且收集函数的调用堆栈信息。调用堆栈是一个动态结构，用于存储正在处理器上执行的函数信息。探查器分析并确定处理器是否正在执行目标进程中的代码。如果处理器没有执行目标进程中的代码，则将放弃样本。

第二种是检测（Instrumentation）。检测模式的工作原理是用 VSInstr 程序在原始的代码中插入一些用于计算时间的代码。例如 A 函数调用 B 函数，那么在调用 B 函数的前后都会被插入用于计算时间的代码。

采样和检测，前者是宏观的性能数据采集，后者是微观的性能数据采集。对于 CPU 负载较高的程序，用采样会得到比较好的效果；但是如果程序运行过程中并没有消耗很多的 CPU 资源，那么用采样可能无法收集到太多有用的信息。所以在不同的场景下需要应用不同的性能数据收集方法。

2.1.2 VSTS 性能分析工具使用

下面以一个简单的例子说一下 Profiler 的用法。示例代码非常简单，新建一个 Website，新建一个页面并在页面中放置一个 Button 控件，然后为 Button 注册 onclick 服务端事件。如下：

```
protected void Button1_Click(object sender, EventArgs e)
{
    ChangedType1();
    ChangedType2();
}

/// <summary>
/// 装箱/拆箱
```

```csharp
/// </summary>
private void ChangedType1()
{
    ArrayList arr = new ArrayList();
    for (int i = 0; i < 10000; i++)
    {
        arr.Add(i);
    }
    int j = 0;
    for (int i = 0; i < 10000; i++)
    {
        j = (int)arr[i];
    }
}

/// <summary>
/// 装箱/拆箱
/// </summary>
private void ChangedType2()
{
    List<int> list = new List<int>();
    for (int i = 0; i < 10000; i++)
    {
        list.Add(i);
    }
    int j = 0;
    for (int i = 0; i < 10000; i++)
    {
        j = list[i];
    }
}
```

代码中的两个方法完成相同的功能。其中 ChangeType1 使用弱类型集合实现；ChangeType2 使用强类型实现。后面将会分析这两个方法哪个执行效率更高。

打开 Visual Studio，单击"分析"→"启用性能向导"选项，如图 2-1 所示。

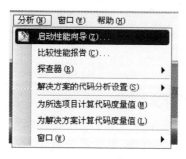

图 2-1 启动性能向导

这里要分析的是 ASP.NET 的程序，在"性能向导--第 1 页（共 4 页）"中，选择"分析 ASP.NET 应用程序"，如图 2-2 所示。

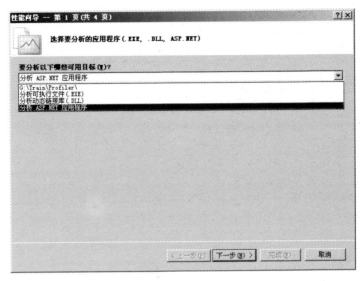

图 2-2　性能向导 1

输入要进行性能分析的站点路径，如图 2-3 所示。

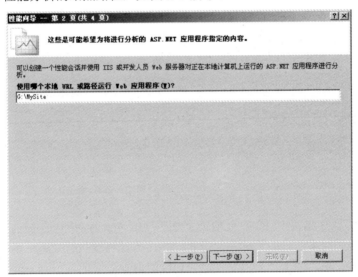

图 2-3　性能向导 2

说明：
- 对于基于服务器（IIS）的网站，请输入 URL，如 http://localhost/MySite/default.aspx。这样将分析本地计算机上位于 MySite 的应用程序根目录下的 ASP.NET 应用程序，同时在 Internet Explorer 中启动该网站上的网页 default.aspx，从而启动会话。
- 对于基于文件的网站，请输入路径，如 G:\MySite\default.aspx。这样将分析位于 G:\MySite 的 ASP.NET 应用程序，同时在 Internet Explorer 中启动网页 http://localhost:nnnn/MySite/default.aspx，从而启动会话。

设置好路径后，单击"下一步"按钮，在"性能向导--第 3 页（共 4 页）"中，选择"检测"分析方法，如图 2-4 所示。

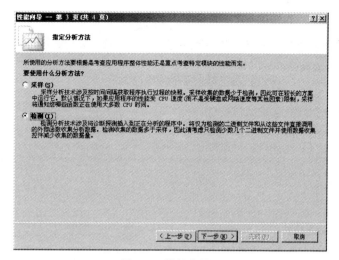

图 2-4　性能向导 3

单击"下一步"按钮。到此为止已经完成一个性能会话的设定，单击"完成"按钮。

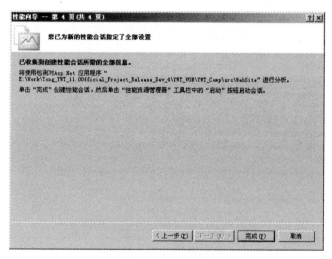

图 2-5　性能向导 4

此时在"性能资源管理器"中，可以看到刚刚建立的性能分析项目，如图 2-6 所示。

图 2-6　创建的性能分析项目

如果看不到此窗口，可以单击"分析"→"窗口"，打开此窗口。此外，在此窗口中通过右击"目标"文件夹，还可以添加目标二进制文件（dll 或 exe）进行分析。如图 2-7 所示。

图 2-7　添加目标二进制文件菜单

到此为止，已经建立好性能分析项目。下面就启动此分析项目，在图 2-8 中有两个选项："启动并启用分析功能"和"启动并暂停分析"。本例并不需要分析 Default.aspx 整个页面，而是仅分析此页面呈现后，按钮事件的执行，因此我们选择"启动并暂停分析"选项。选择"启动并暂停分析"后，即可以启用一个浏览器和 Default.aspx 页面。

图 2-8　启动分析项目

VSTS Profiler 自动启用浏览器后，在开始分析之前，单击"数据收集控件"窗口中的"继续收集"按钮，即可以收集数据，如图 2-9 所示。

图 2-9　单击"继续收集"按钮

如果在"数据收集控件"窗口中看不到，也请单击"分析"→"窗口"打开。切换到浏览器窗口，单击页面中的 Button 控件，则 VSTS Profiler 已经收集了 Button 事件信息。然后，切换到"数据收集控件"窗口，单击"停止"按钮，VSTS Profiler 将停止检测并以报表形式显示收集的信息，如图 2-10 所示。

图 2-10　停止分析命令

报表默认显示摘要视图，如图 2-11 所示。

图 2-11　摘要视图

在摘要视图中，会显示三个列表：最频繁调用的函数、大部分时间单独工作的函数和耗时最长的函数，每个列表默认显示前 5 个。单击上方的"当前视图"下拉框，切换到调用关系视图，如图 2-12 所示。

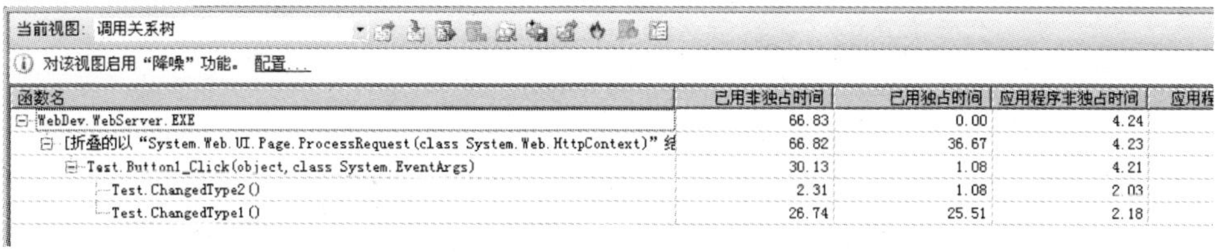

图 2-12　调用关系视图

在此视图中，可以看到方法之间的调用关系。如 Button_Click 事件调用了 ChangeType2 和 ChangeType1 两个方法。同时可以看到，弱类型实现方法 ChangeType1 消耗时间为 26.74ms，而强类型集合对象实现方法 ChangeType2 仅消耗 2.31ms，效率相差 10 多倍。因此在实现集合存储业务类或控件集合属性时，应尽量使用强类型集合。

另外，VSTS Profiler 报表分析器还提供了模块、调用方/被调用方、函数、标记等报表展现形式，可供分析。使用 VSTS Profiler 需要注意以下两点：

（1）VSTS Profiler 不支持对加密的 dll 进行分析，因此在分析时应替换为加密之前的 dll 进行分析。

（2）启用 VSTS Profiler 一直到停止分析之间，不要修改分析站点下面的任何文件。Profiler 在分析时会修改当前分析的所有目标二进制文件及 web.config 等文件，在每个 dll 中的每个方法开始/结束位置插入时间标记并以此计算当前方法执行时间；同时，Profiler 会备份修改前的 dll，在停止分析后用备份的 dll 还原修改过的 dll。例如，如果在此过程中修改了 web.config 文件并保存，则有可能使 VSTS Profiler 工程进程无法正确还原分析之前备份的 web.config。

2.1.3 VSTS 报表字段字典

性能分析时，最重要的一个步骤即是看懂分析器报表，下面将对各种报表所有列的字段的说明见表 2-1。

表 2-1 报表列表字段说明

列　名	说　明
应用程序独占时间百分比	在此上下文的所有函数实例的应用程序独占时间中，指定数据段的总时间（处理器时钟周期数）或性能计数器值所占的百分比
应用程序包含时间百分比	在此上下文的所有函数实例的应用程序非独占时间中，指定数据范围的总时间（处理器时钟周期数）或性能计数器值所占的百分比
已用独占时间百分比	在此上下文的所有函数实例的已用独占时间中，指定数据段的总时间（处理器时钟周期数）或性能计数器值所占的百分比
已用包含时间百分比	在此上下文的所有函数实例的已用非独占时间中，指定数据段的总时间（处理器时钟周期数）或性能计数器值所占的百分比
应用程序	创建进程或线程的应用程序的文件名
应用程序 Etl 文件	包含 ETW 数据的文件的位置
应用程序独占时间	函数或摘要项的计算时间，此时间不包括包含操作系统事件的性能数据、集合探测的时间和函数调用的子例程的时间
应用程序包含时间	函数或摘要项的计算时间，此时间不包括包含操作系统事件的性能数据和集合探测的时间，但包括函数调用的子例程的应用程序时间
平均应用程序独占时间	指定数据范围中所有此函数实例的平均应用程序独占时钟时间（处理器时钟周期数）、P6 性能计数器值或用户定义计数器值
平均应用程序包含时间	指定数据范围中所有函数实例的平均应用程序包含时钟时间（处理器时钟周期数）、P6 性能计数器值或用户定义计数器值
平均已用独占时间	指定数据范围中所有函数实例的平均已用独占时钟时间（处理器时钟周期数）、P6 性能计数器值或用户定义计数器值
平均已用包含时间	指定数据范围中所有函数实例的平均已用包含时钟时间（处理器时钟周期数）、P6 性能计数器值或用户定义计数器值
基址	所加载模块的内存地址
调用深度	调用关系树的深度
时钟频率	处理器的时钟速度

(续)

列 名	说 明
命令行	用于创建性能报告的命令
计数器 1 - 16	检测期间,可以从多个计数器收集数据。默认情况下,仅使用 TimeStamp 计数器
CPU ID	CPU 制造商和类型信息
创建时间	这是创建 .vsp 文件的日期/时间
不正常关机	表示探查器突然退出
已用独占时间	函数或摘要项的计算时间,此时间不包括集合探测的时间以及函数调用的子例程的已用时间
已用包含时间	函数或摘要项的计算时间,此时间包括函数调用的子例程的间隔,但不包括集合探测的时间
事件操作	发生的事件操作的类型。其事件操作可为进入或退出,它仅显示在调用关系树报告中
事件类型	事件操作等事件类型仅显示在调用关系树报告中。下面的列表解释不同的事件类型: L - 加载模块 M - 标记 E - 结尾 Explicit - 指显式接受函数出口 Implicit - 指因为意外而导致函数退出 另外,如果启用了 ETW 事件,则事件类型中也会列出它们
独占分配	一个函数中的分配,不包括该函数调用的子例程中的所有其他分配
独占分配百分比	(一个函数中除了其调用的子例程中的所有其他分配/分析期间发生的独占分配的总数)×100
独占分配字节数	一个函数中分配的字节数,其中不包括该函数调用的子例程中的所有其他字节分配
独占字节百分比	(一个函数中分配的字节数/总字节数)×100
独占百分比	(函数的独占样本/分析期间的总独占样本)×100
独占样本数	为函数收集的性能数据样本的总数,其中不包括该函数调用的其他函数的性能数据
独占转换次数	在此函数(不包括此函数调用的函数)的所有实例中发生转换(操作系统)事件的次数
独占转换百分比	在数据范围内此函数实例的非独占时间内发生,同时在函数的独占时间内发生的操作系统(转换)事件总数的百分比
最终进程数	分析运行结束时处于活动状态的进程数,通常仅在与采样分离时为非零值
最终线程数	应用程序终止时,正在运行的活动已分析的线程数
函数地址	十六进制格式函数的地址
函数名	函数的名称
I/O 缓冲区	用于在分析过程中存储数据的缓冲区数
ID	为进程或线程分配的系统定义的数字标识符
非独占分配	一个函数中的分配,其中包括该函数调用的子例程中的所有分配
包含分配百分比	(一个函数中包含其调用的子例程中的所有分配/分析期间发生的包含分配的总数)×100
包含字节百分比	(一个函数及其调用的所有其他子例程中分配的字节数/总字节数)×100
包含分配字节数	一个函数中分配的字节数,其中包括该函数调用的子例程中的所有其他字节分配
包含字节百分比	(一个函数及其调用的所有其他子例程中分配的字节数/总字节数)×100
非独占百分比	(函数的包含样本/分析期间的总包含样本)×100

(续)

列　名	说　明
非独占样本数	为函数收集的性能数据的总数，其中包括该函数调用的其他函数的性能数据
非独占转换次数	在此函数（包括此函数调用的函数）的所有实例中发生操作系统（转换）事件的次数
包含转换百分比	调用关系树中此函数的父函数调用的此函数实例的非独占时间内发生的操作系统（转换）事件与数据范围中非独占转换总次数的百分比
指令地址	十六进制格式指令的地址
内核 Etl 文件	如果收集了 ETW 数据，会使用这些数据创建一个单独的文件，这是包含内核事件的文件的位置
级别	调用关系树中的调用深度
行号	源文件中函数开始的位置
计算机名	用于分析的计算机
标记	由用户作为标记插入到代码中以帮助记录性能问题的数据
最大应用程序独占时间	调用关系树中父函数调用的所有此函数实例的最大单个应用程序独占时钟时间（处理器时钟周期数）或性能计数器值
最大应用程序包含时间	调用关系树中父函数调用的所有此函数实例的最大应用程序包含时钟时间（处理器时钟周期数）或性能计数器值
最大已用独占时间	调用关系树中父函数调用的所有此函数实例的最大已用独占时钟时间（处理器时钟周期数）或性能计数器值
最大进程数	探查器允许一次分析的最大进程数，超过该最大数将报告错误。可通过注册表项控制该最大数
最大线程数	探查器允许一次分析的最大线程数，超过该数量将报告错误。可通过注册表项控制该最大数
最小应用程序独占时间	调用关系树中父函数调用的所有此函数实例的最小应用程序独占时钟时间（处理器时钟周期数）或性能计数器值
最小应用程序包含时间	调用关系树中父函数调用的所有此函数实例的最小应用程序包含时钟时间（处理器时钟周期数）或性能计数器值
最小已用独占时间	调用关系树中父函数调用的所有此函数实例的最小已用独占时钟时间（处理器时钟周期数）或性能计数器值
最小已用包含时间	调用关系树中父函数调用的所有此函数实例的最小已用包含时钟时间（处理器时钟周期数）或性能计数器值
模块标识符	用于跟踪模块的无符号整数，将根据模块加载到进程中的顺序分配此数字。例如，第一个模块为模块 0，而第二个模块为模块 1 等
模块名	包含函数的模块名称
模块路径	模块的目录位置
模块大小	以十六进制表示的模块大小
名称	使用 NameProfile API 函数分配给进程或线程的字符串。如果没有为项分配名称，则在项 ID 前预置项类型（进程、线程）
CPU 数	用于分析的计算机中的 CPU 总数
调用数	在调用关系树中父函数对此函数实例的调用次数
操作系统	操作系统版本信息
父函数地址	调用另一个函数的父函数在内存中的地址
调用百分比	调用关系树中此函数的父函数调用的此函数的实例数与指定数据范围内所有函数的总调用次数的百分比

（续）

列　名	说　明
进程高位	同时分析的最大进程数
进程 ID	进程的数字标识符
进程名	进程的名称
报告创建时间	创建报告的日期和时间
根节点递归	指示此函数是否在此上下文中被直接或间接地递归调用
样本间隔	所使用的样本或事件之间的平均时钟周期数
样本（已中止）	无法遍历应用程序堆栈时所取的样本
样本（应用程序）	应用程序处于应用程序模式时所取的样本
样本（已损坏）	由于调用堆栈遍历无法进行到线程起始地址导致的不完整堆栈
样本（内核）	应用程序处于内核模式时所取的样本。这些是引发产生的
样本（系统开销）	进行调用堆栈遍历时发生的样本。这些是引发产生的
样本（总数）	发生的样本总数
方案名	默认值为 ProfileRun，可以使用 NameProfile API 在进程中设置
缓冲区大小	I/O 缓冲区的大小，参见"I/O 缓冲区"
源文件	.vsp 报告文件的完整路径
源文件名	包含此函数的源文件的名称
堆栈指针	函数的堆栈指针的值
源字符的开始位置	报告代码样本开始处的列号
源字符的结束位置	报告代码样本结尾处的列号
源行的开始位置	报告代码样本开始处的行号
源行的结束位置	报告代码样本结尾处的行号。这基本与"源行的开始位置"相同，不同之处在于在多行语句上采样，例如： y = myFunction(x + 2, y * 3, x / y) – x;
线程高位	一次分析的最大线程数
线程 ID	为线程分配的系统定义的数字标识符
线程名	使用 NameProfile API 函数分配给进程或线程的字符串。如果没有为项分配名称，则在项 ID 前预置项类型（进程、线程）
时间差	发生此事件的时间戳和前一个事件的时间戳之间的差
时间独占探测系统开销	在调用关系树中父函数调用的此函数实例的独占时间中，分析探测所用的总时间
时间包含探测系统开销	在调用关系树中父函数调用的此函数实例的包含时间中，分析探测所用的总时间
时间戳	发生事件的时间
工具名称和完整版本	VSPerfReport 的名称和版本
进程总数	分析会话生存期内分析的进程总数
线程总数	分析过程中创建的线程总数
类型	表示调用关系树中行的类型。值： 0 表示根函数 1 表示调用方 2 表示被调用方 例如，如果有以下调用关系树： main→funcA→funcB

（续）

列　名	说　明
类型	则类型列将具有以下行： 0 - main 2 - funcA 0 - funcA 1 - main 2 - funcB 0 - funcB 1 - funcA
唯一 ID	标识该函数的十六进制数字
唯一进程 ID	用于跟踪进程的无符号整数。将根据进程加载的顺序分配此数字。例如，第一个进程为 0，而第二个进程为 1 等
VSPerf90.DLL 文件版本	数据收集动态链接库的版本，名为 VSPerf90.DLL，它创建了 .vsp 文件

另外 DotTrace 和 AQTime 也是工作中常用的，在调试不同的应用程序时应选择合适的工具。

2.2　客户端代码性能分析方法

2.2.1　IE8 Profiler 简介

IE8 Profiler 是 IE8 浏览器集成开发人员工具的一个探查器。使用此工具可以探测到当前页面请求过程中所执行的脚本文件的一些信息，如脚本函数名称、函数执行的时间、函数调用的次数、函数所在文件，还可以直接定位到具体的脚本位置查看完整的脚本定义。

要使用 IE8 Profiler，应安装 IE8 浏览器。然后通过单击浏览器菜单中的"工具"→"开发人员工具"，或按〈F12〉快捷键，即可打开 IE8 Profiler。

2.2.2　使用 IE8 Profiler 分析客户端脚本

步骤一：打开需要进行分析的页面，并在浏览器中运行。本示例页面非常简单，页面中代码如下：

```
<input id="Button1" type="button" value="IE Profiler" onclick="KFun1();" />
```

在页面运行时，页面上仅有一个按钮。对应的 KFun1 方法在另一个脚本文件中，如下：

```
function KFun1()
{
    for (val = 0; val < 30000; val++)
    {
        document.getElementById("button1");
    }
    KFun2();
}

function KFun2()
```

```
        {
            KFun3();
        }
        function KFun3()
        {
            alert('执行完成！');
        }
```

共定义三个 JS 函数：KFun1、KFun2、KFun3，其中 KFun1 函数调用 KFun2，KFun2 内部又调用 KFun3 函数。

步骤二：按〈F12〉快捷键或依次单击浏览器菜单中的"工具"→"开发人员工具"，即可打开开发人员集成工具。

步骤三：在开发人员工具窗口中，选择"探查器"选项卡，如图 2-13 所示。然后单击"开始配置文件"按钮，打开探查器的跟踪功能。

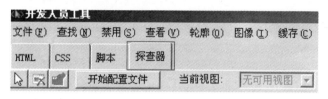

图 2-13　探查器选项卡

步骤四：单击步骤一中已经运行的页面中的按钮。

步骤五：回到开发人员工具窗口，单击"停止配置文件"按钮。即可以看到探查器探查到的所有执行脚本信息，如图 2-14 所示。

图 2-14　探查器报表

以上即为使用 IE8 Profiler 的一个过程，接下来将介绍怎样分析探查器报表。

2.2.3　查看 IE8 Profiler 分析报告

1．脚本跟踪

在探查器分析结果列表中，双击某行即可以进入当前脚本定义。

2．选择视图查看方式

选择"当前视图"下拉框，查看函数视图或调用树视图，如图 2-15 和图 2-16 所示。

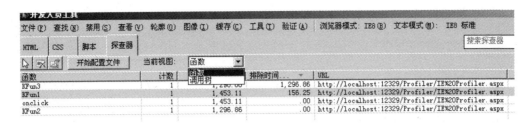

图 2-15　函数视图

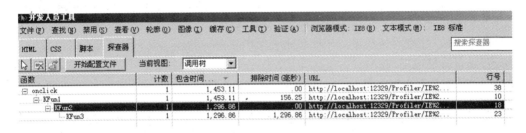

图 2-16　调用树视图

3．配置统计列表列

在统计列表中右击，选择"添加/删除列"即可进行列定义。还可以通过选中列标题进行拖动，改变列的显示位置。

但本设置只对当前有效，下次打开时又恢复为默认系统列状态（如图 2-17 所示即为系统默认显示的所有列）。

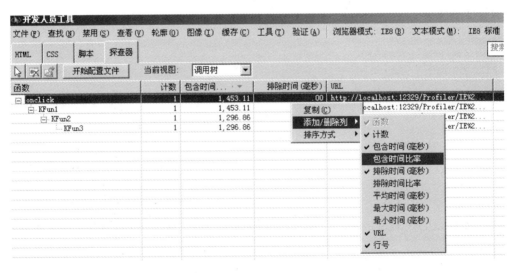

图 2-17　自定义统计列表

4．搜索功能

在探查器右上角有个搜索框，当页面中脚本资源比较多时，可以通过此搜索框直接定位要查找的脚本文件，如图 2-18 所示。

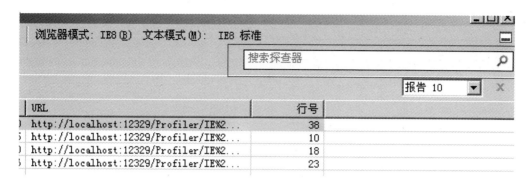

图 2-18　搜索功能

5. 分析报告导出

单击图 2-19 中的“导出”按钮，可以把报表导出为 Excel 文件格式的文件，如图 2-19 所示。

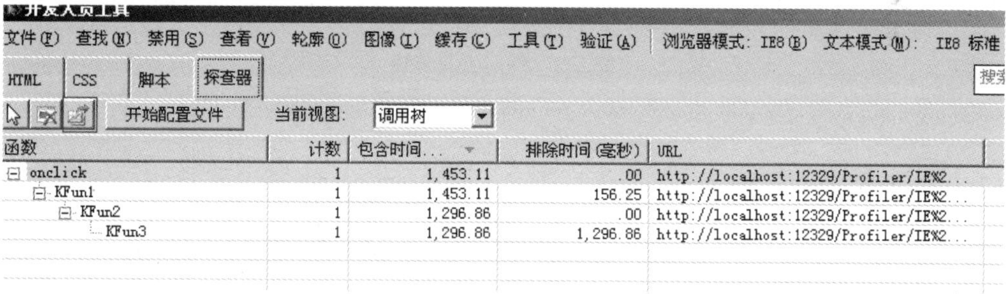

图 2-19　分析报告导出

导出的格式为函数视图，如图 2-20 所示为本示例导出的数据。

	A	B	C	D	E	F	G	H	I	J	K
1	函数	计数	包含时间(	包含时间比	排除时间(	排除时间比	平均时间(	最大时间(	最小时间(	URL	行号
2	KFun3	1	1,296.86	89.25	1,296.86	89.25	1,296.86	1,296.86	1,296.86	http://lc	23
3	KFun1	1	1,453.11	100	156.25	10.75	1,453.11	1,453.11	1,453.11	http://lc	10
4	onclick	1	1,453.11	100	0	0	1,453.11	1,453.11	1,453.11	http://lc	38
5	KFun2	1	1,296.86	89.25	0	0	1,296.86	1,296.86	1,296.86	http://lc	18

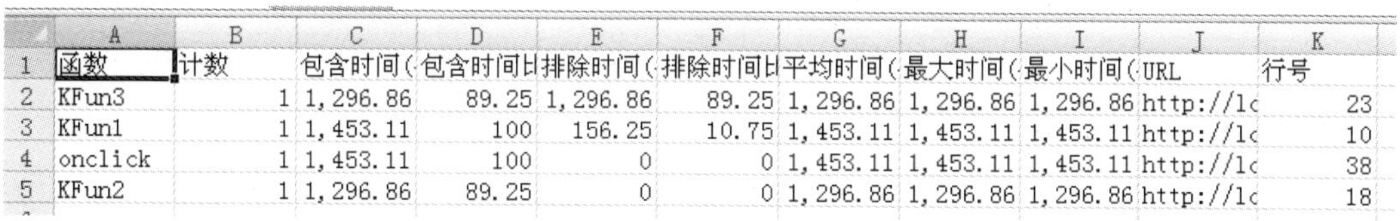

图 2-20　报告导出为 Excel 格式

2.2.4　IE8 Profiler 报表字段字典

与前面介绍的 VSTS Profiler 一样，本节也对 IE8 Profiler 报表字段做全面解释说明。如表 2-2 所示。

表 2-2　报表列表字段说明

列　　名	说　　明
Function	函数名
Count	函数总共被调用次数
Inclusive Time (ms)	函数包括其子函数消耗时间，以毫秒为单位

(续)

列　名	说　明
Inclusive Time %	函数包括其子函数消耗时间百分比
Exclusive Time (ms)	函数本身不包括其子函数消耗时间
Exclusive Time %	函数本身不包括其子函数消耗时间所占百分比
Avg Time (ms)	函数执行平均时间。如果本函数在期间调用了 3 次，则平均时间为调用时间/3
Max Time (ms)	函数包括其子函数消耗最大时间，以毫秒为单位
Min Time (ms)	函数包括其子函数消耗最小时间，以毫秒为单位
URL	函数定义所在的文件地址（Url）
Line Number	函数所在文件中的行数

2.3　性能调优工具集锦

如表 2-3 所示为性能调优的一些工具及简单介绍。

表 2-3　性能工具集锦

工　具	工　具　介　绍
SQL Sever Profiler	SQL 性能优化最常用的一个工具，捕捉事件信息和 SQL 语句
Developer tools	IE 浏览器自带，可以进行 JS 性能分析、JS 调试和 HTML/CSS 布局等操作
HTTPWatch	一款功能强大的数据包抓取软件。获取客户端与服务器交互时的网络数据时，通过此工具可以大概评估问题出在哪个环节，如网络还是后台
CLRProfiler	CLR 内存分析软件
JSLeaker	JS 内存泄漏监控软件，在低版本的浏览器中应用较多
IEJSLeaksDetector	JS 内存泄漏监控软件，专用于 IE
WAST	WEB 服务器压力测试工具
Windows Perfmon	Windows 性能监控工具，能监控计算机系统中所有软件和硬件，在性能调优中非常实用，后面专门有一章进行讲解
VSTSProfiler	单点性能分析工具，对开发人员非常实用
DebugDiag	可以对应用程序系统（C/S 和 B/S）运行异常、内存泄漏等跟踪，与 WinDbg 类似，只是这个工具是可视化的，易上手但不如 WinDbg 灵活
IISState	IIS 服务器工具
TinyGet	一款压力测试工具，模拟并发请求
FireBug	Firefox 下的插件，与 Developer tools 功能类似
Reflector Pro	代码反射工具。在性能调优过程中，大多数情况下是没有代码的
PageDetailer	与 HTTPWatch 功能类似，只是由 IBM alphaworks 发布
IISTrace	IIS 服务器工具
dynaTrace Ajax Edition	前端性能分析工具，除了调试 JS，还可以调试 Ajax
WinDbg	功能很强大的调试软件，由微软发布。可以调试服务器 Crach、线程 hang、CPU High 占用、内存泄漏等任何问题。后面有一章案例讲解中用到此工具较多
Fiddler	与 HTTPWatch 功能类似
.NET Memory Profiler	一款.NET 可视化的内存分析软件，我用它解决过内存泄漏问题

（续）

工　具	工　具　介　绍
Drip/sIEve	内存泄漏检测工具
LoadRunner	大并发压力测试工具。目前最好的一款并发测试工具，功能强大
VSLoadTest	VS 自带的负载测试工具
AQTime	性能、异常等跟踪工具，可用于 C/S、B/S 等多平台下的应用程序
dotTrace	.NET 下性能分析工具
SqlInternalsViewer	数据库页分析工具，可以直接看到数据库文件中存储的页结构和页数据
linux:nmon	Linux 下的性能监控工具，可以监控 CPU、内存、磁盘等运行状况
Dr. Watson	华生医生，应用程序崩溃时，可以由它生成 DUMP 文件
WireShark	一款非常棒的 UNIX 和 Windows 上的开源网络协议分析软件

这些工具比较多，如果要一一介绍恐怕得写一本书，建议在实际工作中边学边用。另外，本书目的主要是讲解性能优化思路，表格中工具的介绍仅仅起引导作用，工具基本用法网络上有很多资料；在后面各个章节中，也都会有这些工具的介绍。

第3章 编码规范

本章内容

- 概述
- 数据库设计及编码规范
- 客户端代码编码规范
- 服务器代码编码规范

3.1 概述

对于一个产品来说，开发人员的接口无非是代码、SQL。如果在开发过程中严格遵守编码规范，则会避免后期出现很多不必要的问题，这对缩短产品开发周期也是很有帮助的。

这些编码规范来源于工作中的不断积累和借鉴，涉及编码的方方面面，很多是平常工作中注意不到的内容，希望对读者有所帮助。

主要内容涉及三个基本方面：数据库设计及编码规范，服务器端代码规范，客户端代码规范。

3.2 编码规范

本节规范以 SQL Server 数据库为例进行说明。

3.2.1 数据库设计及编码规范

1．控制数据库碎片增长过快

碎片增长原因在数据库优化章节的专题研究中，有详细介绍。这里着重强调一下相关代码规范，如果想更详细了解数据库优化内容，请见第 6 章。

以下是笔者在工作中实际遇到的问题和解决方案。

（1）每个表至少保留一个聚集索引

表按存储方式分为聚集表——创建聚集索引的表和非聚集表（堆）——未创建聚集索引的表两种类型。

对于聚集表，可以通过重新创建或重新组织达到对表数据碎片清理的目的，因为聚集索引在完全二叉树的叶节点即是表数据位置。而对于堆，由于没有聚集索引，数据存储相对比较散列，即使碎片整理也只能对所有非聚集索引进行整理，不能对数据本身进行整理。

如果表不含有聚集索引，即使采用 SSMS 的数据库或文件收缩功能，也无法收缩，导致账套太大，不好管理，查询速度变慢。

（2）避免使用有可能被再次修改的字段作为聚集索引

虽然我们建议用"时间"字段作为聚焦索引，但聚集索引字段尽量用 CreatedTime 这种永远不改变的字段代替 updatedTime 这种后续可能被修改的字段。

因为一旦聚集索引字段被修改，由于索引的规则数据库要对字段重新排序，则会导致数据的批量移动、更新效率较低，也容易导致碎片产生。

说明：

CreateTime（创建时间）：在数据生成时写入当前时间作为初始值，后续不可修改。
UpdateTime（修改时间）：数据每被修改一次就更新一次时间，值为当前修改数据的时间。

（3）Guid 类型字段建立索引后会产生较多碎片

◆ 对表的主外键关系（主键和外键）避免使用 Guid 类型标识，一旦使用必然会创建索引。

◆ 对 Guid 类型的字段不建议创建索引。
◆ Guid 类型字段应该禁止建立聚集索引。

由于 Guid 值大小具有随机性，这样的字段建立聚集索引经常会导致即使一个很小的插入也会产生大量批量数据移动。

Guid 除了大小的随机性，它本身占用的字节也比较大，会导致索引用很少时间就产生大量碎片。

如果产品中违反了以上规则，并且已经在使用了，可以通过如下方式进行优化。

1）数据库端修正

SQL Server 也有内置的函数，即 NEWSEQUENTIALID()。目前默认的函数为 NEWGUID()，是无序的。这样可以有效避免在插入数据时索引碎片的产生。

2）代码端生成修正

根据 Guid+服务器网卡 ID+时间，自己生成有序的 Guid。

对 Guid 类型对碎片的影响，同样举以下几个例子。

首先，对比 int 和 Guid 对碎片影响，插入 10 万条数据后，碎片情况如表 3-1 所示。

表 3-1　磁盘碎片情况（1）

	表名	表数据量	索引名	索引类型	页填充度	索引碎片	页数
☑	T1	99999	I_ID	NONCLUSTERED INDEX	65.899629355...	99.011857707...	506
☑	T2	99999	I_ID	NONCLUSTERED INDEX	99.610032122...	2.1505376344...	186

Guid 类型的索引碎片约为 99%，而 int 类型的索引碎片仅为 2%左右。

再对比一下有序生成 Guid 和无序生成 Guid 对碎片的影响。

同样，插入 10 万条数据，碎片情况如表 3-2 所示。

表 3-2　磁盘碎片情况（2）

	表名	表数据量	索引名	索引类型	页填充度	索引碎片	页数
☑	T1	19998	I_ID	NONCLUSTERED INDEX	68.470471954...	99.009900990...	101
☑	T2	19998	I_ID	NONCLUSTERED INDEX	97.412107734...	52.112676056...	71

执行随机插入操作后，表 T1 和 T2 的碎片情况有了明显的差异，有序 Guid 索引产生的碎片大概是无序 Guid 索引碎片的一半。

在数据空间占用方面，采用有序 Guid 方式仅占用了 71 页，而无序 Guid 方式占用了 101 页。

（4）尽量用系统的基本类型表示相应的业务字段类型

有时候为了表示统一，一些业务字段不用系统基本类型表示。例如，某个产品中没有用 DateTime 基本日期类型字段表示日期，而是用了 char(20)表示日期，其实验数据如表 3-3 所示。

表 3-3　实验数据

	表名	表数据量	索引名	索引类型	页填充度	索引碎片	页数
☑	T2	99999	I_CreateTime2	CLUSTERED INDEX	61.795997034...	69.292742050...	14026
☑	T1	99999	I_CreateTime2	CLUSTERED INDEX	94.561984185...	0.6159938400...	9091

顺序插入 10 万条数据后，聚集索引列类型为 char(19)的索引碎片增加到 69%；而 DateTime 列的索引并没有增加，说明 char(19)类型列作为索引会导致大量碎片的产生。

因此要尽量用系统的基本类型表示相应的业务字段类型。

（5）数据库碎片归纳总结

顺序插入数据时，一般数据类型，如 int, DateTime 等不会产生碎片；Char(19)这样的类型会产生较多的碎片；Guid 类型更容易产生碎片，除了长度，更主要的原因是它产生值的随机性；索引列字段占字节越大，产生的索引碎片越大。

随机插入数据时，所有数据类型都易产生碎片；char(19)数据字段索引比基本类型（int、DateTime 等）更易产生碎片；聚集索引比非聚集索引更易产生碎片。

随机更新时，更新聚集索引列消耗的时间更长；无论是聚集表还是堆表随机更新比随机插入产生的碎片更多；更新聚集索引比更新非聚集索引代价更大，因为更新聚集索引，会对所有非聚集索引进行更新，而更新非聚集索引则仅对这一个索引进行修改。

执行删除操作对碎片影响非常小，基本可以忽略不计。

2．索引优化

（1）索引创建规则

- ◆ 为经常被用于查询的谓词创建索引，如下拉参照中被用于快速查找的 code 和 name 等。在平台现有下拉参照的查询 SQL 语句中，like 的条件语句要取消前置通配符。还需要关注 Order By 和 Group By 谓词的索引设计，Order By 和 Group By 的谓词是需要排序的，某些情况下为 Order By 和 Group By 的谓词建立索引，会忽略查询时的排序动作。
- ◆ 对于内容存在大量重复的列，比如只包含 1 和 0，应禁止建立索引。因为该索引选择性极差，在特定状况下会误导优化器做出错误的选择，导致查询速度大幅下降。
- ◆ 当一个索引由多个列构成时，应注意将选择性强的列放在前面。前后次序稍有不同，性能就可能出现数量级的差异。
- ◆ 对小表进行索引有时并不能产生优化效果，因为查询优化器在遍历用于搜索数据的索引时，花费的时间可能比扫描简单的表还长，因此设计索引时需要考虑表的大小。对于记录数不大于 100 的表禁止建立索引，频繁操作的小数据量表不建议建立索引（记录数不大于 3000）。

（2）聚集索引创建的必要性

实验表明，一旦创建聚集索引后再删除聚集索引，查询计划会由聚集搜索变为原来的表扫描，但扫描成本仍然为 15%，而不是之前的 41%。换句话说，由于聚集索引能够保证数据物理存储是连续的，所以一般查找大数据量时会比较快，磁盘轴每次物理寻道（移动）或逻辑 lookup/rid 操作时间较短；如果表中没有聚集索引，则可能数据是分散存储的，用一般的表扫描会使磁盘轴每次寻道（移动）或逻辑 lookup/rid 操作的时间比较长，整个表扫描时间就会随之增长很多。

聚集索引可以避免表中产生数据碎片，建议在表中保留一个聚集索引。产生数据碎片的主要原因是表中缺少聚集索引，如果表中含有聚集索引可以使数据在任何更新操作（插入/删除）后自动移动数据以保证数据的物理连续，很少会产生数据存储间的空闲碎片。因为聚集索引结构中的叶索引就是数据位置，而这些是非聚集索引做不到的。解决办法是，要在表中保留一个聚集索引，这样在更新（插入/修改/删除）时就可以尽量避免数据碎片的产生，而不是等到账套大了再缩小。那些不含有聚集索引且插入更新频繁的表是最容易导致体积变大的。建议使用 CreateTime，用户一般是按时间插入单据的，按索引排序原则会插入到数据最后，这样也不会导致数据匹配移动；尽量不要用 ID(Guid)，由于 Guid 值的大小具有随机性，这样的字段建立聚集索引经常会发生即使一个很小的插入也会导致大量批量数据移动的情况。用 CreateTime 作聚集索引也便于在日后数据量很大时进行数据分区。

最重要一点，虽然大部分时候聚集索引扫描与表扫描的查询速度是一样的，但聚集索引查找是所有操作中最快的，比非聚集索引查找还快。

（3）索引类型的选择

在不同情况下，该创建聚集索引还是非聚集索引，如图 3-1 所示。

动作描述	使用聚集索引	使用非聚集索引
列经常被分组排序	√	√
返回某范围内的数据	√	×
一个或极少不同值	×	×
小数目的不同值	√	×
大数目的不同值	×	√
频繁更新的列	×	√
外键列	√	√
主键列	√	√
频繁修改索引列	×	√

图 3-1　创建索引类型建议

3．避免阻塞设计原则

（1）整体原则

总结起来性能优化也依赖于代码和 SQL 的优化，逻辑代码要优化到最快，否则一个耗时较长的 SQL 语句将会阻塞全部用户等待几十秒甚至几分钟，所以针对查询 SQL 语句的优化是最重要的。

修改批量操作的需求，产品中单记录操作的消耗不小，批量操作耗时和记录数量成正比。因此，设计时要避免在一个自动事务服务方法中做批量的循环操作，这里可以将循环操作放到 UI 控制器端，这样就使一个长事务变成多个短小的事务，减少阻塞机会。

减少写操作使用锁的数量，比如减少批量更新操作中修改行的数量，使行锁定减少，同时减少锁升级至表锁的概率。

一些耗时长、锁定数据多的操作需要避免和正常业务操作冲突，可以使用调度计划在系统闲置的时间内运行，或者使用互斥机制来保证其他用户暂时退出操作以独立运行，具体视业务情况而定。

（2）处理原则

1）减少读操作需要的共享锁

a）将事务隔离级别由默认读提交（ReadCommited）修改为读未提交（ReadUn Commited），这样将不会有读/写阻塞，但是会造成读取其他事务未提交的数据。指定 Isolation Level 属性为 ReadUn Commited。

b）在 SELECT 语句中带 NoLock 提示，事务内无锁提示也不会有读/写阻塞，与上面一样也会有脏读。SELECT 上加无锁控制，在执行 SELECT 操作时加 with(NoLock)，如：SELECT * FROM person WITH(NoLock)。

这两种方式都会出现脏读，对数据要求非常严格的场景要慎用。

2）在 SQL Server 中使用基于行版本的快照隔离模式

从 SQL Server 2005 版本开始，数据库支持行版本快照隔离功能，主要是用于解决读/写阻塞问题，提供了以下两种方式。

◆ 方式一：已提交读快照打开

ALTER DATABASE 数据库名 SET READ_COMMITTED_SNAPSHOT ON;

◆ 方式二：快照隔离打开

ALTER DATABASE 数据库名 SET ALLOW_SNAPSHOT_ISOLATION ON;
[SET TRANSACTION ISOLATION LEVEL SNAPSHOT]

方式一和方式二在使用时略有不同。方式二除了要打开快照隔离，在每个事务中使用时都要配合 SET TRANSACTION ISOLATION LEVEL SNAPSHOT 子句使用才能生效；而方式一只要打开快照即可生效。方式一主要是针对物理对象（如表、行、键、索引等），方式二针对逻辑事务，所以方式一控制粒度更小。

4．死锁最小化原则

（1）死锁产生原因

死锁是由两个互相阻塞的线程组成的，它们互相等待对方完成。一般死锁情况下，两个数据库事务之间存在着反向的操作。SQL Server 中死锁监视器定期检查死锁，如果发现死锁，将选择其中回滚消耗最小的任务，这时易发生 1205 数据库错误。

SQL Server 有几十种锁，以下是两个最典型的：

共享锁（S 锁）：如果事务 T 对数据 A 加上共享锁，则其他事务只能对 A 再加共享锁，不能加排他锁。获准共享锁的事务只能读取数据，不能修改数据。

排他锁（X 锁）：如果事务 T 对数据 A 加上排他锁，则其他事务不能再对 A 加任何类型的锁。获准排他锁的事务既能读取数据，又能修改数据。

（2）死锁最小化

1）写/写死锁

用相同的顺序访问对象，如果涉及多于一张表的操作，要保证事务中都按照相同的顺序访问这些表。

另外，要减少一个事务中大批量更新的操作，大批量的写操作涉及记录独占锁太多且一直到事务结束才能释放，更容易与其他事务造成死锁。

2）读/写死锁

去掉读操作共享锁的最佳方式是使用 SQL Server 的快照模式，其次是使用读取未提交隔离模式或使用 with no lock 提示。

5．高性能 SQL 建议

（1）确保查询索引有效

1）索引使用原则

创建了索引，并不能确保 SQL 语句在执行时一定用到。以下是能有效使用索引的条件语句：

> [fieldname]=8
> [fieldname]>200
> [fieldname] between 0 and 100
> [fieldname] like 'abc%'

以下是不能有效利用索引的一些语句：

> SUM（[fieldname]）=8
> [fieldname] + 1 =8
> [fieldname] like '%abc%'

2）避免对表达式左侧索引字段进行运算或函数处理

低效的写法为：

> select name from table where substring(name,1,3) = 'king'

高效的写法为：

> select name from table where name like 'king%'

3）复合索引使用原则
- 选择性强的条件放在前面。
- 尽量用复合索引，速度快而且不会导致表中索引个数太多，还能节省空间。

（2）高效率 SQL 语句

1）查询时不返回不需要的行和列

在查询时，尽量使用通配符，如 SELECT * FROM T1 语句，要用到几列就选择几列，如 SELECT C1,C2 FROM T1。在可能的情况下尽量限制结果集行数，如 SELECT TOP 100 C1,C2,C3 FROM T1，因为某些情况下用户是不需要那么多数据的。

2）合理使用 EXISTS,NOT EXISTS 子句

> SELECT SUM(T1.C1)FROM T1 WHERE((SELECT COUNT(*) FROM T2 WHERE T2.C2=T1.C2>0)
> SELECT SUM(T1.C1) FROM T1 WHERE EXISTS(SELECT * FROM T2 WHERE T2.C2=T1.C2)

以上两种写法的结果相同，但是后者的效率显然要高于前者。因为后者不会产生大量锁定的表扫描或索引扫描。

还有以下几个 SQL 语句：

> SELECT a.hdr_key　FROM hdr_tbl a　　　//tbl a 表示 tbl 用别名 a 代替
> WHERE NOT EXISTS (SELECT * FROM dtl_tbl b WHERE a.hdr_key = b.hdr_key)
>
> SELECT a.hdr_key　FROM hdr_tbl a
> LEFT JOIN dtl_tbl b ON a.hdr_key = b.hdr_key　WHERE b.hdr_key IS NULL
>
> SELECT hdr_key　FROM hdr_tbl
> WHERE hdr_key NOT IN (SELECT hdr_key FROM dtl_tbl)

以上三种写法都可以得到同样正确的结果，但是效率依次降低。
3）最大化使用可能的查询条件
例如，以下两个 SQL 中第二句很可能比第一句执行快得多。

  SELECT SUM(A.Money)
  FROM T1 A left join T2 B on A.ID = B.ID

  SELECT SUM(A.Money)
  FROM T1 A left join T2 B on A.ID = B.ID　　AND
    A.CODE=B.CODE

4）使用 UNION ALL 代替 UNION
合并两个结果集时，尽量使用 UNION ALL，因为它在执行的时候不会进行"确定唯一排序"，而 UNION 则要排序。
如果对结果集不需要排序，或者数据肯定是唯一的，则要用 UNION ALL 代替 UNION 合并运算符。SQL 语句如下：

  select c1, c2, c3 from t1 where code = '001'
  union all
  select c1, c2, c3 from t1 where code = '002'

肯定要比以下 SQL 语句高效。

  select c1, c2, c3 from t1 where code = '001'
  union
  select c1, c2, c3 from t1 where code = '002'

3.2.2　客户端代码编码规范

下面这些规范以 JavaScript 语言为例进行说明。

1. 预先计算 elements 的长度

JavaScript 与服务端代码一样，不要重复计算当前遍历集合元素的长度。
低效的写法如图 3-2 所示：
高效的写法如图 3-3 所示：

```
function C1 ()
{
    var ements = document.getElementsByTagName("SPAN") :
    var index = 0:
    while(index < 1000 )
    {
        for( var i = 0: i < ements.length:  i++ )
        {
            var a = i.toString():
        }
        index++:
    }
}
```

```
function C2 ()
{
    var ements = document.getElementsByTagName("SPAN") :
    var index = 0:
    while(index < 1000 )
    {
        for( var i = 0, k=ements.length:i<k: i++ )
        {
            var a = i.toString():
        }
        index++:
    }
}
```

  图 3-2　低效写法   图 3-3　高效写法

分别用两种方式访问 collection.length 属性，实验统计结果如图 3-4 所示。

函数	计数	包含时间（毫秒）
C1	1	124.99
C2	1	46.87

图 3-4　统计结果

2．高效率变量的定义方式

低效的写法为：

```
function Compute()
{
    var   i;
    var   j;
    var   k;
    for(i=0; i < X.Length; i++)
        for(j=0; j < Y.Length; j++)
            for(k=0; k < Z.Length; k++)
            {
                ……（省略完整代码）
            }
}
```

高效的写法为：

```
function Compute()
{
    for(var i=0; i < X.Length; i++)
        for(var j=0; j < Y.Length; j++)
            for(var k=0; k < Z.Length; k++)
            {
                ……（省略完整代码）
            }
}
```

再如，低效的写法为：

```
p = new Person();
p.Name = "King";
p.Age = "20";
p.Sex = "male";
```

高效的写法为：

```
p = { Name: "King", Age: "20", Sex: "male" }
```

JavaScript 是解释执行语言，并且每条代码都会作为一个单位进行提交执行，因此要减少提交的次数，尽量用最少的语句实现相同的功能。

3．对象缓存优化

由于 JavaScript 的解释性，a.b.c.d.e 需要进行至少 4 次查询操作，先检查 a 再检查 a 中的 b，再检查 b 中的 c，等等。

请看图 3-5 中的示例，这种写法每次访问元素都要挂上"document.all…"。如图 3-6 所示为高效率写法，这种写法是先把 document.all 和 document.all.length 存储到变量中，使用时直接访问变量，避免再次查询的过程。

```
function Q1()
{
    var n=0;
    while(3000)
    {
        for(Var i=0;i< document.all.length; i++)
        {
            if( document.all(i).tagName =="SPAN")
            {
                a+= document.all(i).id ;
            }
        }
        n++;
        if(n>10000)break;
    }
}
```

```
function Q2()
{
    var n=0;
    while(3000)
    {
        var all=document.all;
        var len=all.length;
        for(var i=0;i<len;i++)
        {
            if( all(i).tagName =="SPAN")
            {
                a+= all(i).id ;
            }
        }
        n++;
        if(n>10000)break;
    }
}
```

图 3-5　低效写法　　　　　　　　　　图 3-6　高效写法

执行 document.all 检索大约 15 万次，两种方式时间消耗差 3 倍，实验统计数据如图 3-7 所示。

Q1	1	15,073.97
Q2	1	5,128.90

图 3-7　统计结果

4．字符串连接高效率写法

如果要连接多个字符串，应该使用"+"，如：

　　s+=a;
　　s+=b;
　　s+=c;

应该写成：

　　s+=a + b + c;

连接大量的字符串，应使用 Array 的 join 方法。
如果是收集字符串，最好使用 JavaScript 数组缓存，最后使用 join 方法连接起来，如下：

```
var array = new Array();
for(var I = 0; I < 100; i++)
{
    array.push(i.toString());
}
var result = array.join("");
```

看一个实验例子，低效的写法如图 3-8 所示，高效的写法如图 3-9 所示。
两种字符串连接方式各执行 1000000 次，时间统计数据如图 3-10 所示。
两种写法时间相差 100ms 左右。

```
function AppendString1()
{
    var a="ABCDEFGHIJKLMNOPQRSTUVWXYZZ1";
    var b="ABCDEFGHIJKLMNOPQRSTUVWXYZZ2";
    var c="ABCDEFGHIJKLMNOPQRSTUVWXYZZ2";
    var s2;
    for(var i=0;i<1000000;i++)
    {
        s2+=a;
        s2+=b;
        s2+=c;
    }
}
```

图 3-8 低效写法

```
function AppendString2()
{
    var a="ABCDEFGHIJKLMNOPQRSTUVWXYZZ4";
    var b="ABCDEFGHIJKLMNOPQRSTUVWXYZZ5";
    var c="ABCDEFGHIJKLMNOPQRSTUVWXYZZ6";
    var s3;
    for(var i=0;i<1000000;i++)
    {
        s3+=a+b+c;
    }
}
```

图 3-9 高效写法

函数	计数	包含时间（毫秒）
AppendString1	1	1,312.47
AppendString2	1	1,249.97

图 3-10 统计结果

再看一个实验，高效的写法如图 3-11 所示，低效的写法如图 3-12 所示。

```
function AppendBigString1()
{
    var buffer=new Array();
    for(var i=0;i<100000;i++)
    {
        buffer.push(i);
    }
    var s=buffer.join('-');
    //alert(s);
}
```

图 3-11 高效写法

```
function AppendBigString2()
{
    var buffer=new Array();
    var s;
    for (var i=0;i<100000;i++)
    {
        s+=i.toString()+'-';
    }
    //alert(s);
}
```

图 3-12 低效写法

两种方式（使用 Array.join 和串连接符）各自执行 100000 次，时间统计数据如图 3-13 所示。

函数	计数	包含时间（毫秒）
AppendBigString1	1	125.00
AppendBigString2	1	187.50

图 3-13 统计结果

5．类型转换函数

使用 Math.floor()或者 Math.round()将浮点数转换成整型。

浮点数转换成整型很容易出错，很多人喜欢使用 parseInt()，其实 parseInt()适用于将字符串转换成数字，而不是浮点数和整型之间的转换。如果是纯数值之间的转换，应该使用 Math.floor() 或者 Math.round()。

再看下面的实验数据，低效的写法如图 3-14 所示，高效的写法如图 3-15 所示。

```
function ChangeToInt1()
{
    var r;
    for(var i=0;I<100000;i++)
    {
        r= parseInt(i);
    }
}
```

图 3-14 低效写法

```
function ChangeToInt2()
{
    var r;
    for(var i=0;I<100000;i++)
    {
        r= Math.floor(i);
    }
}
```

图 3-15 高效写法

使用 parseInt()和 Math.floor()两种方法分别转换 100000 次，统计数据如图 3-16 所示

函数	计数	包含时间（毫秒）
ChangeToInt1	1	93.75
ChangeToInt2	1	62.50

图 3-16　统计结果

6．避免闭包写法

使用 prototype 代替 closure，可以有效地避免闭包。使用 closure 在性能和内存消耗上都是不利的，如果 closure 使用量过大，这就会成为一个问题。所以，尽量将：

```
this.method1 = function()
{
    ……（省略完整代码）
}
```

替换成：

```
Class1.protoype.methodFoo = function()
{
    ……（省略完整代码）
}
```

和 closure 存在于对象实例之中不同，prototype 存在于类中，被该类的所有的对象实例共享，可以代码重用。

另外，闭包在有些浏览器（或一些浏览器版本）中会出现内存泄漏，需要特别注意。

7．高效操作 DOM——离线操作

在添加一个复杂的 DOM 树时，可以先构造，构造结束后再将其添加到 DOM 树的适当节点；如果操作已经填加到节点的 DOM，可以先移除(脱离)，构造完成后再添加到节点上。
看一下以下代码：

```
var ul = document.getElementById("ul");
for(var i=0; i < 100; i++)
{
    ul.appendChild(document.createElement("li"));
}
```

下面这种写法会更高效：

```
var ul = document.getElementById("ul");
var li = document.createElement ("li");
var parent = ul.parentNode;
parent.removeChild(ul);
for(var i=0; i<100; i++)
{
    ul.appendChild(li.cloneNode(true));
}
parent.appendChild(ul);
```

同样做了一个实验,数据如下。一般写法如图 3-17 所示,更高效的写法如图 3-18 所示。

```
function CreateDom1()
{
    var divParent=document.createElement("DIV");
    document.body.appendChild(divParent);
    for(var ;=0;I<10000;i++)
    {
        var span=document.createElement("SPAN");
        divParent.appendChild(span);
    }
}
```

图 3-17　低效写法

```
function CreateDom2 ()
{
    var divParent=document.createElement("DIV");
    for(var ;=0;I<10000;i++)
    {
        var span=document.createElement("SPAN");
        divParent.appendChild(span);
    }
    document.body.appendChild(divParent);
}
```

图 3-18　高效写法

构造完后再追加到 DOM 元素上,防止单次构造操作导致浏览器回流操作。

在上述实验中,用两种方式分别创建 10000 个 SPAN,并追加到 body 中,执行时间比较如图 3-19 所示。

函数	计数	包含时间(毫秒)
CreateDom1	1	234.37
CreateDom2	1	203.12

图 3-19　统计结果

3.2.3　服务器代码编码规范

本节规范以 C#语言为例进行说明。

1. 预先计算 collection 的 length

不要在循环语句中循环计算当前集合遍历长度。

如将

```
for(var i=0; i < collection.length; i++)
{
    ……(省略完整代码)
wq}
```

替换成

```
for(var i=0; len=collection.length; i<len; i++)
{
    ……(省略完整代码)
}
```

效果会更好,尤其是在大循环中。

2. 避免循环体内重复创建对象

避免循环体内重复创建对象的代码如下。

```
List< BusinessObj > Compute(Collection sets)
{
    List< BusinessObj > list = new List<Results>();
    foreach(RelateObj obj in set.Length)
    {
```

```
                Creater creater = new Creater();
                BusinessObj obj = creater.Create(obj);
                list.Add(obj);
        }
        return list;
}
```

高效的做法是将 builder 对象提到循环外面创建。

3．优先使用系统常量

看一下这段代码有没有优化的余地？

```
If(object1.Amount == new decimal(0))
{
        Object1.Status = AmountStatus.None;
}
```

正确的做法是使用 Decimal.Zero 常量，使用常量避免创建对象。

4．使用 StringBuilder 做字符串连接

第 1 章中第二个案例就是因为 "+"（String.Concat 方法）连接导致了 GC 99% CPU 利用率。当连接字符串个数超过 10 个时要用 StringBuilder，如 StringBuilder sb = new StringBuilder(**256**)。

另外，后面的参数为 256，StringBuilder 类默认申请 Buffer 的长度为 16，一旦超出这个范围 Buffer 就会重新分配，这比较消耗资源。在我们已知要连接字符串长度的情况下，要传一个初始值参数，这样首次就会分配指定的空间大小。

5．避免不必要的调用 ToUpper 或 ToLower 方法

```
Pubic bool GetTrueOrFalse
{
        Get
        {
                return bool.Parse(object1.ToString().ToLower());
        }
}
```

这里的 ToLower 方法完全没必要。

6．用 String.Compare 方法代替 ToLower 和==运算

请看如下代码：

```
Foreach(XmlElements element in a.Elements)
{
        If(element.Value.ToLower()== "a")
        {
                … …
        }
        If(element.Value.ToLower()== "b")
        {
                … …
```

```
                }
                If(element.Value.ToLower()== "c")
                {
                    ……（省略完整代码）
                }
            }
```

高效的做法是使用 Compare 方法，这个方法可以做大小写忽略的比较，并且不会创建新字符串，如下：

```
If(String.compare(element.Value, "a", StringComparison.OrdinalIgnoreCase) == 0)
```

7．多线程代码优化

（1）线程同步：同步方法

以下是多线程控制同步的实现方法。

```
[MethodImpl(MethodImplOptions.Synchronized)]
public static PerformanManager GetInstance()
{
    if(instance == null)
    {
        PM pm    = new PerfmonManager();
    }
    return pm;
}
```

请注意：如果能控制同步语句尽量同步语句，并且遵循同步粒度最小原则。

（2）线程同步：同步类型

锁定 Type 对象会影响同一进程中所有应用程序域的所有实例，这不仅可能导致严重的性能问题，也会导致各种错误的发生。应额外构造一个 static 的成员变量，让此类成员变量作为锁定对象。

```
public static PerformanManager GetInstance()
{
    if(instance == null)
    {
        lock(typeof(Manager))
        {
            PM pm    = new PerfmonManager();
        }
    }
    return pm;
}
```

（3）线程同步：同步 this 对象

要避免锁定 this 对象，因为锁定 this 会影响该实例的所有方法。假设对象 obj 有 A 和 B 两个方法，其中 A 方法使用 lock(this)对方法中的某段代码设置同步保护。现在，因为某种原因，B 方法也开始使用 lock(this)来设置同步保护，并且可能为了完全不同的目的。这样，A 方法就被干扰了，其行为可能无法预知。所以，基于良好的习惯，建议避免使用 lock(this)这种方式。

（4）线程同步：同步策略

采用静态对象的方式实现同步，代码如下。

```
public class UserOnlineStatusManager
{
    private Hashtable htOnlineUsers = null;
    private static Object _lockObject = new Object();
    public static UserOnlineStatusManager GetInstance()
    {
        lock(_lockObject)
        {
            if (instance == null)
            {
                instance = new UserOnlineStatusManager();
            }
        }
        return instance;
    }
}
```

这里把同步对象_lockObject 定义为静态的原因在于，所有线程都要能够对同步对象_lockObject 进行访问才能实现同步功能，也就是_lockObject 对象起到线程之间数据传递作用，而线程之间共享数据常用的方法之一就是通过静态变量，如果定义成非静态变量则每个线程都会有自己的一个_lockObject，也就无法实现同步功能。

（5）线程同步：Double Check 机制

以上代码其实还不够严谨，在某些情况下还是会出现问题。

一般在 lock 语句后就会直接创建对象了，但这不够安全。因为在 lock 锁定对象之前，可能已经有多个线程进入第一个 if 语句中了。如果不加第二个 if 语句，则单例对象会被重复创建，新的实例替代旧的实例。如果单例对象中已有数据不允许被破坏或者存在其他原因，则应考虑使用 Double Check 技术。修正后的代码如下：

```
public class UserOnlineStatusManager
{
    private Hashtable htOnlineUsers = null;
    private static UserOnlineStatusManager instance = null;
    public static UserOnlineStatusManager GetInstance()
    {
        if (instance == null)
        {
            lock(_lockObject)
            {
```

```
            if (instance == null)
            {
                instance = new UserOnlineStatusManager();
            }
        }
    }
    return instance;
}
```

8. 使用泛型集合避免装箱/拆箱

这里直接看实例。

低效的写法如图 3-20 所示。

高效的写法如图 3-21 所示。

```
private void ChangedType1()
{
    ArrayList arr=new ArrayList();
    for(int i=0;i<10000;i++)
    {
        arr.Add(i);
    }
    int j=0;
    for(int i=0;i<10000;i++)
    {
        j=(int)arr[i];
    }
}
```

图 3-20 低效写法

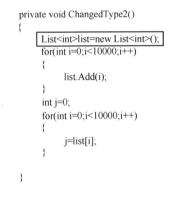

图 3-21 高效写法

分别用非泛型集合和泛型集合对 int 类型的数据各装箱和拆箱 10000 次，得出的比较数据（ms）如图 3-22 所示。

图 3-22 统计结果

9. 避免两次检索集合元素

获取集合元素时，有时需要检查元素是否存在。通常的做法是先调用 ContainsKey（或 Contains），然后再获取集合元素，这种写法非常符合逻辑。

以下这段代码有没有优化的余地？

```
if(HashTable.Exist(Key))
{
    return HashTable[key];
}
```

如果考虑效率，可以先直接获取对象，然后判断对象是否为 null 来确定元素是否存在。对于 Hashtable，则可以节省一次 GetHashCode 调用和 n 次 Equals 比较，避免两次检索集合元素。如下：

```
object value = HashTable[key];
if(value != null)
{
    ……（省略完整代码）//code logic
}
```

10．避免两次类型转换

请看一下以下这段代码有没有优化的余地？

```
if(obj is Employee)
{
    Employee employee = (Employee)obj;
    employee.Do();
}
```

效率更高的写法如下（避免两次类型转换）：

```
Employee employee = obj as Employee;
if(employee != null)
{
    employee.Do();
}
```

第4章
服务器性能监控

本章内容

- 概述
- 服务器性能监控

4.1 概述

一个网站运行在服务器环境下,除了优化网站本身的软件代码外,还要对服务器的软件和硬件进行监控。比如当系统崩溃或响应缓慢时,要跟踪服务器并判断是哪个部件的瓶颈,如 CPU、内存、磁盘、网络等。本章就介绍一下服务器一些常用的性能参数。

在 Windows 操作系统下监控服务器性能参数非常简单,只需要在"开始"→"运行"框中输入"perfmon"命令即可打开"性能"窗口,如图 4-1 所示。

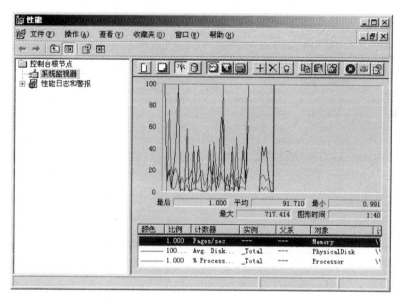

图 4-1　性能窗口

在图 4-1 中,上方显示各个计数器的负载曲线图,下方是计数器列表。Windows 提供了几百个计数器,通过单击工具栏中的"+"命令可以打开"添加计数器"窗口,如图 4-2 所示。

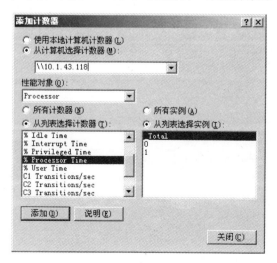

图 4-2　添加计数器窗口

在"添加计数器"窗口中即可以添加想要监控的计数器。Windows 计数器可以支持前台实时监控和后台日志文件记录两种方式,具体要根据应用场景来选择。

此外,LoadRunner 也对服务器计数器监控提供了支持,如图 4-3 所示为 LoadRunner 中的监控窗口。

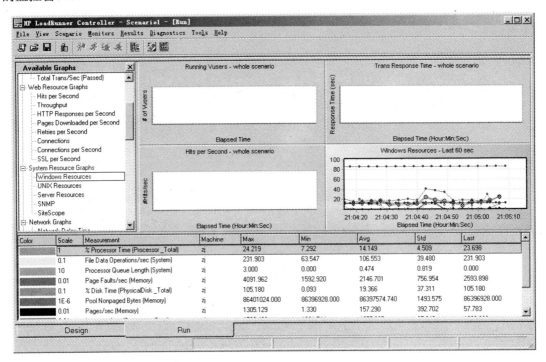

图 4-3 Loadrunner 服务器监控窗口

添加计数器的方式与 Windows 计数器方式相同。

4.2 服务器性能监控

Windows 提供了几百个计数器,在 MSDN 上都有详细的解释,限于篇幅这里就不全部列举出来了。本节主要介绍一些常用的计数器,如果读者没有使用过 Windows 计数器功能,本节可以起引导入门作用。

4.2.1 内存

● Available MBytes

此值表示当前计算机剩余可用物理内存数,以 MB 为单位。系统正常运行时,系统可用物理内存不低于总内存的 10%(或不小于 100MB),否则就需要检测是否代码内存泄漏或添加物理内存硬件来解决。还有两个计数器 Available KBytes 和 Available Bytes,与 Available Mbytes 表示的意义一样,一般对内存进行细粒度监控时使用。

● Committed Bytes

表示当前所有进程提交的内存量。Available Mbytes 和 Committed Bytes 相加之和基本为系统总内存。

- %Committed Bytes in Use

表示正在使用的内存与总内存的比值。如果比值大于 90%，则表示内存紧张。

- Pages/sec

表明由于硬件页面错误而从磁盘取出的页面数，或由于页面错误而写入磁盘以释放工作集空间的页面数。一般 Pages/sec 范围值在 20 以内，如果持续高于 100，那么应该进一步研究页交换活动，确认是否是内存问题。

- Page Reads/sec

页的硬故障，Page/sec 的子集，为了解析对内存的引用，必须读取页文件的次数。阈值为>5，越低越好，大数值表示磁盘读取而不是缓存读取。

在研究内存不足不太明显的页交换的原因时，必须跟踪如下的磁盘使用情况计数器和内存计数器：Physical Disk\ % Disk Time、Physical Disk\ Avg.Disk Queue Length，例如，包括 Page Reads/sec 和 % Disk Time 及 Avg.Disk Queue Length。如果页面读取操作速率很低，同时 % Disk Time 和 Avg.Disk Queue Length 的值很高，则可能有磁盘瓶颈。但是，如果队列长度增加的同时页面读取速率并未降低，则内存不足。

- Pages Input/sec

是为了解决硬错误页从硬盘上读取的**页数**，而 Page Reads/sec 是为了解决硬错误从硬盘读取的**次数**。如果 Page Reads/sec 比率持续保持为 5，表示可能内存不足。

- Page Faults/sec

当处理器在内存中读取某一页出现错误时，就会产生缺页中断，也就是 Page Fault。如果这个页位于内存的其他位置，这种错误称为软错误；如果这个页位于硬盘上，必须从硬盘重新读取，这个错误称为硬错误，硬错误会使系统的运行效率很快降下来。Page Faults/sec 计数器就表示每秒处理的错误页数，包括硬错误和软错误。此计数器正常值在 1000 以内。

4.2.2 处理器

- %Processor Time

表示 CPU 工作时间占总时间的百分比。如果该值持续超过 95%，表明瓶颈是 CPU。可以考虑增加一个处理器或换一个更快的处理器。潜在因素包括内存不足、数据库低查询计划重用率和未经优化的查询。

- %User Time

应用程序占用 CPU 时间的百分比，如数据库排序、执行 Aggregate Functions 等操作，%UserTime 与%Privileged Time 累加值等于%Processor Time。

- %Privileged Time

表示 CPU 内核时间，是在特权模式下处理线程执行代码所花时间的百分比。

- %DPC Time

处理器在网络处理上消耗的时间，该值越低越好。在多处理器系统中，如果这个值大于 50%，并且%Processor Time 非常高，加入一个网卡可能会提高性能。

4.2.3 磁盘

- %Disk Time

指所选磁盘驱动器忙于为读取或写入请求提供服务所用时间的百分比。

如果此值持续超过 90%，则表示磁盘已经产生瓶颈；如果 Disk Time 和 Avg.Disk Queue Length 的值很高，而 Page Reads/sec 页面读取操作速率很低，则可能存在磁盘瓶颈。

- Avg Disk Queue Length

指磁盘读写队列的平均长度，该值应不超过磁盘数的 1.5～2 倍。要提高性能，可增加磁盘。

注意：一个 RAID Disk 实际有多个磁盘。如果使用 RAID 设备，%Disk Time 计数器显示的值可以大于 100%。如果大于 100%，则使用 Avg. Disk Queue Length 计数器决定正在等待磁盘访问的系统请求的平均数。

- Avg Disk Read/Write Queue Length

指读取/写入请求列队的平均数。当 Avg Disk Queue Length 较大时可以分别通过读/写队列长度计数器确认是磁盘读瓶颈还是磁盘写产生了瓶颈。

- Average Disk sec/Read

指以秒计算的在此盘上读取数据所需的平均时间。一般来说，定义该值小于 15ms 为最优，介于 15～30ms 之间为良好，30～60ms 之间为可以接受，超过 60ms 则需要考虑更换硬盘或硬盘的 RAID 方式了。

- Disk Bytes/sec

提供磁盘系统的吞吐率。

4.2.4 网络

- Bytes Received/sec

网卡每秒接收的字节数，即下行。

- Bytes Sent/sec

网卡每秒发送的字节数，即上行。

- Bytes Total/sec

网卡每秒传输的字节数，即下行和上行之和。

在客户现场进行网络诊断是较频繁的工作，在这里就多做一些说明。网速检测有以下两种方法。

方法一：使用网速检测工具。

我最常用的是 DU Meter 或 360 的网速监视器。这些工具能够检测出当前计算机哪些进程各自占用了多少流量，且同时支持上行和下行速度检测。如图 4-4 所示为 Du Meter 网速监视器。

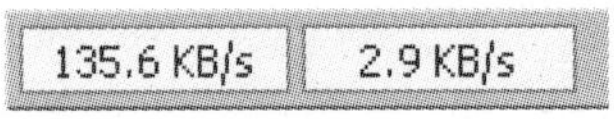

图 4-4　DU Meter 网速监视窗口

可以看到当前下行速度为135.6KB/s（左边值），上行速度为2.9KB/s（右边值）。上行速度监视值用于从本机上传文件到其他服务器时。

我在客户现场使用最多的是DU Meter，因为DU Meter网速检测值比较准确。

方法二：使用Perfmon网络接口对象。

单击"开始"→"运行"，输入命令："perfmon"，则会弹出性能窗口，然后在右侧面板空白处单击鼠标右键，选择"添加计数器"，打开"添加计数器"对话框，如图4-5所示。

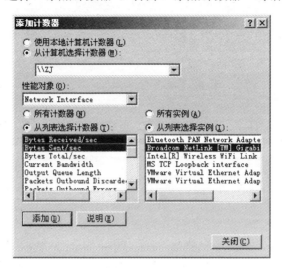

图4-5　DU Meter网速监视窗口

在"性能对象"下拉框中选择"Network Interface"对象，在"从列表选择计数器"列表中选择Bytes Received/sec和Bytes Sent/sec，在"从列表选择实例"列表中选择"Broadcom NetLink [TM] Gigabit Etherent"（可以选择当前网络接口设备），即可将这两个计数器添加到监视窗口，如图4-6所示。

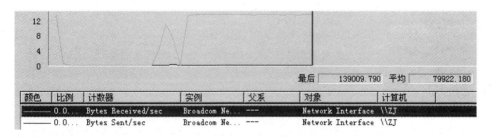

图4-6　Perfmon—Network Interface

可以看到当前下行速度约为139KB/s。

方法三：使用任务管理器中的"联网"功能

打开"Windows任务管理器"，选择"联网"选项卡。在上方会显示所有网络设备网络带宽利用率图形，下方会显示各个网络接口设备的速度，如图4-7所示。

每间隔发送的字节数即为当前上行速度，每间隔接收的字节数即为当前下行速度。如图4-7所示，当前下行速度约为137KB/s。

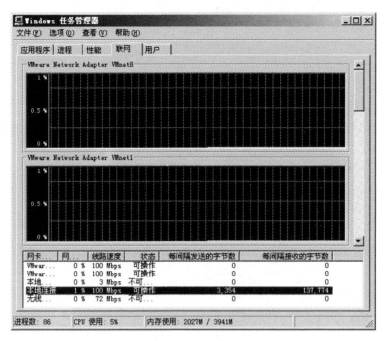

图 4-7 任务管理器窗口

默认情况下，列"每间隔发送的字节数"和"每间隔接收的字节数"是不显示的，需要从任务管理器菜单单击"查看"→"选择列…"中设置。

实际应用中，针对客户现场具体情况，选择上述几种诊断方法中任意一种即可，它们统计的数据都是一样的，只是工具不同。

4.2.5 进程

- %Processor Time

表示进程占用处理器的时间。

- Page Faults/sec

与内存对象下的 Page Faults/sec 相同，只是表示当前进程而不是所有进程导致的页故障。常用于确定是哪个进程导致的页故障。

- Private Bytes

表示进程独占内存，包括物理内存和虚拟内存。如果 Process\Private Bytes 计数器和 Process\Working Set 计数器的值持续升高，同时 Memory\Available Bytes 计数器的值却持续降低，则很有可能发生了内存泄漏。

- Virtual Bytes

Virtual Bytes 指处理器使用的虚拟地址空间以字节数显示的当前大小。

4.2.6 系统

- Processor Queue Length

处理器队列的线程数量。与磁盘计数器不同，此计数器只显示就绪线程，而不是正在运行的线程。因此，如果一台计算机有几个处理器，则需要将此值除以处理工作量的处理器数量。每个处理器，不管工作量大小，如果保持 10 个以下线程的处理器队列，通常是可以接受的。

- Context Switches/sec

表示所有工作线程的切换次数。一般阈值不要超过 5000×CPU 个数。

4.2.7 .NET CLR Memory

- Bytes in All Heaps

指示当前.NET 堆大小。Bytes in All Heaps 记录了上次 GC 发生时所统计到的进程中不能被回收的所有 CLR Object 占用的内存空间。与 Process/Private Bytes 比较可以区分出是 Managed Heap 还是 Native Memory 导致了内存泄漏。

- %Time in GC

表示 GC 发生的频繁程度。一般来说 15%以内算比较正常，如果超过 20%说明 GC 发生过于频繁。由于 GC 不仅仅带来很高的 CPU 开销，还需要挂起目标进程的 CLR 线程，所以高频率 GC 是非常危险的。

4.2.8 .NET CLR Loading

- Current Assemblies

此计数器显示在此应用程序的所有 AppDomain 中加载的程序集的当前数目。程序集太多会导致内存碎片过多。

4.2.9 Asp.net

- Request/sec

每秒钟到达 ASP.NET 的请求数。通过此值可以衡量 ASP.NET 负载的程度。还有一个计数器 Anonymous Request/sec，表示每秒匿名请求数，有匿名请求时可以观察此计数器值。这两个计数器在一般情况下值相等。

- Request Failed

失败的请求数。

- Request in Application Queue

当 ASP.NET 没有空余的工作线程来处理新进入的请求时，新的请求会被放到 Application Queue 中。当 Application Queue 堆积的请求也超过设定数值的时候，ASP.NET 直接返回 503 Server too busy 错误，同时丢弃该请求。正常情况下，Request in Application Queue 应该为 0，否则说明有请求队列等待。

4.2.10 数据库

数据库计数器支持也非常丰富，但只能提供产品级的一些概括统计，如果想精确查明问题产生的原因，还要通过 SQL Server 提供的专门性能调优工具，如 DMV 查询等。

另外，数据库计数器一般没有明确的阈值，因为不同的系统阈值范围都不尽相同，一般要为自己的系统建立一套基数曲线，在性能调优过程中根据现场与此基数曲线的偏差程度来确定问题。

- Access Method
  - √ Full Scans/sec

    每秒不受限的完全扫描数，可以是基本表扫描或全索引扫描。如果这个计数器显示的值较高，应该分析查询以确定是否确实需要全表扫描，优化 SQL 语句。
  - √ Page Splits/sec

    每秒产生的页分割数。由于数据更新操作引起的每秒页分割的数量。
- Buffer Manage
  - √ Buffer Cache Hit Ratio

    可在高速缓存中找到而不需要从磁盘中读取的页的百分比。计数器值依应用程序而定，但比率最好为 90% 或更高。增加内存直到这一数值持续高于 90%，表示 90% 以上的数据请求可以从数据缓冲区中获得所需数据。
  - √ Lazy Writes/sec

    惰性写进程每秒写的缓冲区的数量，值最好为 0。
  - √ Page Life Expectancy

    没有引用的页停留在缓冲池中的时间（秒）。此值越大越好，一般情况下要大于 300（5min），否则可能内存不足。
- Cache Manage
  - √ Cache Hit Ratio

    高速缓存命中次数和查找次数之间的比率，此值一般不低于 80%。
- Locks
  - √ Number of Deadlocks/sec

    每秒钟导致死锁的请求数量，正常情况下为 0。
  - √ Lock Requests/sec

    锁管理器每秒请求的新锁和锁转换数。有些系统几千次，有些系统则几十万次，具体要依据当前系统的基数据曲线。
  - √ Lock Waits/sec

    每秒要求调用者等待的锁请求数。
- SQL Statistics
  - √ Batch Requests/sec

    每秒收到的 Transact-SQL 命令批数，批请求数值高意味着吞吐量很大。
  - √ Compilations/sec

    表示每秒钟编译的 SQL 数目，与 Batch Requests/sec 进行比较。计划重用 =（批处理请求数-编译数）/批处理请求数，比值越大越好。
  - √ Re-Compilations /sec

    表示每秒钟重新进行编译的 SQL 数目。如果该值较大，则需要检查 SQL 语句查询计划未被缓存的原因。
- Memory Manager
  - √ Target Server Memory

    目标服务器内存。即使我们给 SQL Server 设置了最大内存，但受操作系统及内存限制，SQL Server 实际并不能获取我们设置的最大内存，从这个计数器中有可能会发现很多问题。
- √ Total Server Memory

当前已使用的内存，小于等于 Target Server Memory 值。

这两个内存计数器，一般结合 Memory 对象下的可用内存分析。（详细分析方法，请见第一章中第一个内存案例）。

第5章
客户实战案例

本章内容

- 概述
- WinDbg 工具介绍
- 客户问题诊断案例

5.1 概述

WinDbg 是微软发布的一款调试工具，可以用于内核模式和用户模式调试。除了可以调试代码，还可以用来解决服务器出现的各种异常问题。

在实际应用中，WinDbg 解决了涉及 CPU High、内存、性能、异常等方面的问题，作为最强大的性能优化工具之一，它还包括了其他同质工具的部分功能，如用 MemoryProfiler、DebugDialog、CLRProfiler 能解决的一些问题，用 WinDbg 也能够分析解决。

本章中，首先介绍一下 WinDbg 的常用用法，然后通过一些客户实战案例分享使用它解决问题的思路。

本章的这些案例涉及内存泄漏、堆栈溢出导致崩溃、CPU High、EventHander 泄漏、异常检测及 Session 使用建议，并尽量用这些案例来覆盖各种不同的故障以说明尽可能多的问题。

5.2 WinDbg 工具介绍

5.2.1 环境配置

首先安装 WinDbg 软件，为方便使用，作以下常用环境变量配置。

1. 新增系统环境变量

_NT_DEBUGGER_EXTENSION_PATH=C:\WINDOWS\Microsoft.NET\Framework\版本号
_NT_SYMBOL_PATH=SRV*c:\Symbols*http://msdl.microsoft.com/download/symbols

在 Path 系统环境变量中添加：

C:\WINDOWS\Microsoft.NET\Framework\版本号
C:\Program Files\Debugging Tools for Windows (x86)

添加好后，如图 5-1 所示。

图 5-1 添加环境变量

2．Dump 文件或附加进程

WinDbg 或其他工具可以把当前运行进程镜像及信息线索写到一个文件中,这个文件就是 Dump 文件。Dump 文件的好处是可以在本地分析,且不影响服务器运行,但有可能会遇到 Dump 中版本与本地环境版本不一致的情况,要注意这个问题,否则无法调试。

WinDbg 也支持附加到当前运行进程去调试,好处是调试起来方便、快捷。通过命令选择直接附加到进程还是调试已经 Dump 的文件,如图 5-2 所示。

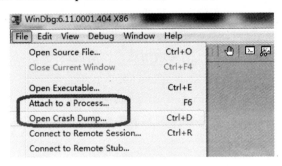

图 5-2　附加进程或打开 Dump 文件

3．调试进程或 Dump 文件

加载好 Dump 文件或附加到进程后,如果是非托管应用进程就可以直接分析了,如果是.NET 应用程序进程,那么微软还提供了 sos.dll 的支持。sos.dll 支持很多命令,更方便调试。它包含在.NET Framework 的安装包中,与.NET Framework 在同一个目录下,不需要单独安装。

4．加载 SOS 扩展功能 dll

　　　.load sos.dll　　　//有时候需要指定 sos.dll 的路径
　　　.loadby sos mscorwks

查看版本,确认是否加载成功。

　　　!eeversion

5.2.2　常用命令简介

得到 Dump 文件:

　　　adplus -hang -pn w3wp.exe -quiet (hang)　　　[-FullOnFirst]
　　　adplus -crash -pn w3wp.exe (crash)

这是两个 Dump 文件的常用命令,在命令窗口中执行。具体用法如下。

　　C:\Program Files\Debugging Tools for Windows (x86) adplus.vbs -hang -o C:\dump -p 2976
　　参数说明:
　　-hang:表示附加到目标进程,抓取 dump 镜像,然后解除。对应的参数是 -crash 崩溃模式,该参数会终止目标进程。
　　-o:指定 Dump 文件保存路径。
　　-p:指定目标进程 PID。

查看异常：

　　!dumpallexceptions

看了以上几个命令，是不是感觉有些困惑呢？这里可以使用 help 指令作为辅助。
运行 help 命令。

　　!help
　　!help [command name]　　//查找某个命令更多详细的内容，如：!help eeversion

显示系统与时间有关的信息：

　　.time

此命令可以得到 Dump 文件时 CPU 的使用情况，还得到其他一些有用的信息，例如等待请求队列的数量、已完成的线程和时间等。

　　!threadpool

列出当前正在运行的线程和 CPU 使用情况。

　　!runaway　　　　　　　　//比如查看哪个线程占用 CPU 时间过多

列表显示出应用程序所有正在运行的线程，当前应用程序域中后台正在运行的程序，以及线程的相关内容等：

　　!threads

切换到特定的线程上查看相关内容：

　　~[thread id]s　　//例如：~2s

列出当前线程中的调用栈信息：

　　!clrstack　　　　//如果想查看额外的信息，添加"-p"选项，如!clrstack –p
　　!clrstack -a　　　//除了显示堆栈还可以显示参数的值（内存地址）

显示指定地址对象的内容：

　　!dumpobj 0xf32fcc　　　　　　　　//!do 为简写

继续查看对象的值：

　　!dumpobj -v 0xf32fcc

　　!dumpvc 7910c878　01573774　　　//mt value

查看当前特定线程的堆栈中所有托管对象：

　　!dumpstackobjects (简称!dso)

得到当前线程对象中关于数组对象的详细信息

```
!dumparray [address]         (简称!da)
!da –details [address]                      //详细信息
```

得到对象的整个大小：

```
!objsize 071bef70
!objsize poi(0x61b47d4+0xc)              //Address + Offset
.foreach (obj {!dumpHeap -mt 0x0c2eaeb4 -short}){!objsize ${obj}}    //0x0c2eaeb4: mt
.foreach(myobj {!dumpHeap -short -min 85000}) {!objsize myobj}       //所有对象中大于 85KB 的内容
```

Dump 出所有托管堆上的对象：

```
!dumpHeap –stat               //使用 -stat 参数来得到托管堆上的摘要信息
```

参数-mt（即 MethodTable）：

```
!dumpHeap -mt 793308ec
```

参数-type：

```
!dumpHeap -type System        //得出许多包含类型为 System 名称的对象
```

参数-min/-max，该参数接受一个最小/最大的对象字节数：

```
!dumpHeap -stat -min 85000                //查看大于 85000bytes 字节的对象
!dumpHeap -min 85000
!dumpHeap -mt 790fd8c4 -min 20000 -max 25000
!dumpHeap -type System.String -min 150 -max 200    //检查大小在 150～200 之间的所有字符串
!dumpHeap -mt 790fd8c4 -strings           //只输出字符串
```

参数-short：

```
!dumpHeap -type System.String -min 6500 –short    //仅查询对象的地址信息
```

.foreach 语句：

```
.foreach (myAddr {!dumpHeap -type System.String -min 6500 -short}){!dumpobj myAddr;.echo ************************}          // .echo 命令打印分割符
```

.shell 命令：

```
.shell -i - -ci "!iisinfo.clientconns" FIND /c "Request active"
.shell -i - -ci "!iisinfo.clientconns" FIND /c "<table"
```

查看进程程序集加载情况：

```
!dumpdomain
!dumpdomain [程序域地址]
!dumpassembly [程序集地址]
```

显示模块信息：

!dumpmodule [-mt] 1c5a1098   //1c5a1098: module address

查看托管线程：

!threads

查看所有的 Exception：

!dumpHeap –type Exception

打印当前堆栈上正在被抛出的 Exception：

!pe/!PrintException

打印出相应命令的详细信息及堆栈：

!pe address

查看某个具体异常的详细信息：

!do [address]

查看对象的信息：

!gcroot [address]

打印出所有同一个类型的 Exception 的信息：

.foreach(myVariable {!dumpHeap -type System.ArgumentNullException -short}){!pe myVariable;.echo **}

列出了 GC 堆的大小和 G0、G1、G2 和 LOH 的开始地址：

!eeHeap [-gc] [-loader]
!eeHeap
!eeHeap -gc
!eeHeap -loader

5.2.3 示例应用

以下是我在学习各种资料时积累的一些常用场景命令，在解决相关的问题时可以查找本节内容。

1. 缓存对象

（1）检查缓存大小

!dumpHeap -stat -type System.Web.Caching.Cache   //得到 System.Web.Caching.Cache 对象的方法表
!dumpHeap -mt 1230494c   //得到 1230494c 方法表中所有对象
!objsize 03392d20   //得到当前地址对应对象的大小

（2）查看什么内容被缓存了

!dumpHeap -stat -type System.Web.Caching   //查看 CacheEntry 对象
!dumpHeap -mt 12306320   //查看 CacheEntry 的方法表 mt

!do 076b42dc   //检查对象的所有内容,之后检查堆栈信息

2. 挂起

检查本地的堆栈信息:

~* kb 2000

检查 dotnet 堆栈信息:

~* e!clrstack

看看有多少调用堆栈里有 Monitor.Enter:

.shell -ci "~* e !clrstack" FIND /C Monitor.Enter   //跟踪诊断挂起现象

检查等待锁的线程 ID 列表:

!syncblk [index]   //提示:MonitorHeld = 1 代表拥有者,2 为等待者

查看一个等待线程的状态:

~5s   //切换到线程 5,用真实的线程 ID 替换 5 即可
kb 2000   //检查本地堆栈信息
!clrstack   //查看.NET 堆栈信息
!clrstack -p   //查看.NET 堆栈信息,包括参数内存地址
!clrstack -a

3. 查看缓存占用情况

查看 Cache 占用内存情况:

!name2ee System.Web.dll System.Web.Caching.Cache
  0:000> !name2ee System.Web.dll System.Web.Caching.Cache
  Module: 65f21000 (System.Web.dll)
  Token: 0x020000fa
  MethodTable: 66148d24
  EEClass: 65f86838
  Name: System.Web.Caching.Cache

!dumpHeap -mt [MethodTable]   //查看托管堆中对象类型

!objsize 06952248   //查看对象大小

4. 内存调试

查看 GC Heap 堆情况:

0:001> !eeHeap -gc
generation 0 starts at 0x0110be64
generation 1 starts at 0x01109cd8
generation 2 starts at 0x01021028
  segment   begin   allocated   size

```
01020000 01021028 0110de70 000ece48(970312)
Total Size    0xece48(970312)
------------------------------
large block 0x11e1fc04(300022788)
large_np_objects start at 17b90008
large_p_objects start at 02020008
------------------------------
GC Heap Size 0x11f0ca4c(300993100)
```

在这里 GC 堆是 300MB 左右，Dump 文件是 358MB。使用 !dumpHeap -stat 命令来查看占用了空间的托管对象：

```
0:001> !dumpHeap -stat
Bad MethodTable for Obj at 0110d2a4
Last good object: 0110d280
total 14459 objects
Statistics:
      MT       Count TotalSize Class Name
3c6185c          1         12 System.Web.UI.ValidatorCollection
3c2e110          1         12 System.Web.Configuration.MachineKeyConfigHandler
3c29778          1         12 System.Web.Configuration.HttpCapabilitiesSectionHandler
3c23240          1         12 System.Web.SafeStringResource
… …
D12f28        1133      46052 System.Object[]
153cb0          88      76216 Free
321b278         85     178972 System.Byte[]
d141b0        6612     416720 System.String
Total 14459 objects
```

使用!gcroot 16220018 从 LOH 上得到更多对象的信息：

```
0:001> !gcroot 16220018
Scan Thread 1 (4e8)
Scan Thread 5 (bb0)
Scan Thread 6 (d0)
Scan Thread 10 (43c)
Scan Thread 11 (308)
Scan Thread 12 (6e4)
Scan HandleTable 14e340
Scan HandleTable 150e40
Scan HandleTable 1a6fa8
HANDLE(Strong):37411d8:Root:020784d8(System.Object[])-
>0108b504(System.Web.HttpRuntime)->0108b9d0(System.Web.Caching.CacheSingle)-
>0108ca68(System.Web.Caching.CacheUsage)->0108ca78(System.Object[])-
>0108cb3c(System.Web.Caching.UsageBucket)-
>010f95fc(System.Web.Caching.UsageEntry[])-
>01109be8 (System.Web.Caching.CacheEntry)->00000000()
```

要找出 System.Web.Caching.Cache 的地址，可使用!name2ee 命令，这个命令接受 2 个参数，即程序集名和全类名：

```
0:001> !name2ee System.Web.dll System.Web.Caching.Cache
--------------------------------------
MethodTable: 03887998
EEClass: 03768814
Name: System.Web.Caching.Cache
--------------------------------------
```

EEClass 是一个用来表示.NET 类的内部结构。
取得托管堆中的某个对象的类型，使用!dumpHeap – mt MethodTable 地址的方式来获得：

```
0:001> !dumpHeap -mt 03887998
  Address        MT      Size
 0108b8ac  03887998        12
Bad MethodTable for Obj at 0110d2a4
Last good object: 0110d280
total 1 objects
Statistics:
      MT    Count TotalSize Class Name
 3887998        1        12 System.Web.Caching.Cache
Total 1 objects
large objects
  Address        MT      Size
total 0 large objects
```

查看 System.Web.Caching.Cache 的大小，使用 !objsize 0108b8ac：

```
0:001> !objsize 0108b8ac
sizeof(0108b8ac) = 300126128 (0x11e38fb0) bytes (System.Web.Caching.Cache)
```

查看托管堆的大小：

```
0:000> !eeHeap -gc
generation 0 starts at 0x012cc0e4
generation 1 starts at 0x012afde8
generation 2 starts at 0x011c1028
 segment     begin  allocated     size
011c0000  011c1028  012d6000  00114fd8(1134552)
Total Size 0x114fd8(1134552)
------------------------------
large block 0x8060(32864)
large_np_objects start at 00000000
large_p_objects start at 021c0008
------------------------------
GC Heap Size 0x11d038(1167416)
```

Dump 显示 GC 堆的大小是 1MB。

5. 线程状态

使用!threads 命令查看当前 CLR 中有哪些线程正在执行：

```
0:004> !threads
ThreadCount: 2
UnstartedThread: 0
BackgroundThread: 1
PendingThread: 0
DeadThread: 0
                    PreEmptive  GC Alloc         Lock
   ID ThreadOBJ    State   GC   Context          Domain  Count APT Exception
0  6ec 0014e708    6020 Enabled 00000000:00000000 00148a90    0 STA
2  a68 00157618    b220 Enabled 00000000:00000000 00148a90    0 MTA (Finalizer)
```

6. 调试.NET 代码

（1）!name2ee SimpleSample.exe SimpleSample.Program.Main 显示方法相关地址：

```
0:004> !name2ee SimpleSample.exe SimpleSample.Program.Main
Module: 00982c5c (SimpleSample.exe)
Token: 0x06000005
MethodDesc: 00983000
Name: SimpleSample.Program.Main()
JITTED Code Address: 01220070
```

（2）!dumpil 00983000 显示方法被 C#编译器编译之后的 IL 代码：

```
0:004> !dumpil 00983000
ilAddr = 004020c4
IL_0000: nop
IL_0001: ldstr "Any key continue... ... "
IL_0006: call System.Console::WriteLine
IL_000b: nop
IL_000c: call System.Console::Read
IL_0011: pop
IL_0012: call SimpleSample.Program::getcharBuffer
IL_0017: stloc.0
IL_0018: ldloc.0
IL_0019: call SimpleSample.Program::changeto4p
IL_001e: nop
IL_001f: ldloc.0
IL_0020: call System.Console::WriteLine
IL_0025: nop
IL_0026: call System.Console::Read
IL_002b: pop
IL_002c: call System.Console::Read
IL_0031: pop
```

IL_0032: ret

（3）!u 01220070 显示 JIT 编译了的方法的本地代码：

Other:
　　!dumpmt -md 00983024　　　　//得到类成员的函数详细信息
　　!dumpHeap -stat　　　　　　　//显示程序中所有对象的统计信息
　　!dumpHeap -mt 00983024　　　//该命令显示 MethodTable 的详细信息
　　!gcroot 012919b8　　　　　　 //来显示一个实例的所属关系
　　!dumpobj(do) 012a3904　　　　//显示一个对象的具体内容
　　!ObjSize 012a1ba4　　　　　　//对象实际在内存中的大小
　　!DumpArray　　　　　　　　　　//查看数组信息
　　!dumpHeap -type Exception　　//查看异常信息

7. 查看方法代码

　　!ip2md 05600dfd　　　　　　　//05600dfd 表示 EIP
　　　　MethodDesc: 02429048
　　　　Method Name: DataLayer.GetFeaturedProducts()
　　　　Class: 055b18ac
　　　　MethodTable: 0242905c
　　　　mdToken: 06000008
　　　　Module: 024285cc
　　　　IsJitted: yes
　　　　m_CodeOrIL: 05600dd0

（1）根据 md 来看：!dumpil 02429048 这个地址是上面步骤 f 中的输出的第一行 MethodDesc 的值。

（2）根据 native code 来看：!u 05600dd0 这个地址是上面步骤 f 中的输出的最后一行的 m_CodeOrIL 的值。

（3）根据 module 来看：!dumpmodule 024285cc 这个地址是上面执行结果的 Module 的值。

5.3 客户问题诊断案例

在前面讲解性能优化思路时分享了两个案例，下面再分享几个案例，通过这些实战案例来说明遇到性能问题时的定位过程。

5.3.1 Web 服务器内存达到 3GB 后崩溃原因诊断定位

1. 问题描述

客户的 Web 服务器运行过程中，IIS 主进程 w3wp 占用内存达到近 3GB，随后系统崩溃（一般情况下该进程占用内存基本在 600MB 以下）。重启机器后，10min 之内又出现 IIS 进程达到 3GB，系统再次崩溃，可见重启也无法解决此问题。而内存达到 3GB，是这个产品之前从来没有遇到过的。

除此之外，客户还提供了一个线索：在任务管理器中他们看到所有物理内存已经被用完，如图 5-3 所示。

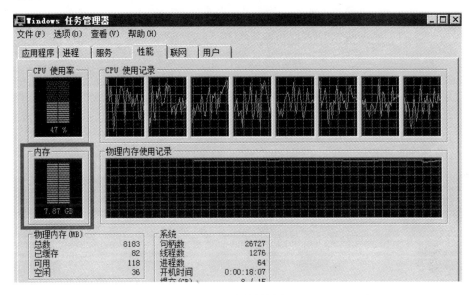

图 5-3　内存被用完

在如图 5-3 所示的任务管理器中，还可以看到 CPU 运行正常，没有负载。监控一下 IIS 内存一系列参数如图 5-4 所示。

最新 2,656,389,244　平均 2,475,618,517　最小 1,769,550,428

显示	颜色	比例	计数器	实例	父系	对象	计算机
☑		0.001	Lock Wait Time (ms)	_Total	---	SQLServer:Locks	\\WIN-P88LTF4SMII
☑		0.0001	Target Server Memory ...	---	---	SQLServer:Memory Manager	\\WIN-P88LTF4SMII
☑		0.0001	Total Server Memory (KB)	---	---	SQLServer:Memory Manager	\\WIN-P88LTF4SMII
☑		0.000001	# Bytes in all Heaps	w3wp	---	.NET CLR Memory	\\WIN-P88LTF4SMII
☑		0.1	# GC Handles	w3wp	---	.NET CLR Memory	\\WIN-P88LTF4SMII
☑		0.1	# Gen 0 Collections	w3wp	---	.NET CLR Memory	\\WIN-P88LTF4SMII
☑		0.1	# Gen 1 Collections	w3wp	---	.NET CLR Memory	\\WIN-P88LTF4SMII
☑		0.1	# Gen 2 Collections	w3wp	---	.NET CLR Memory	\\WIN-P88LTF4SMII
☑		0.000001	# Total committed Bytes	w3wp	---	.NET CLR Memory	\\WIN-P88LTF4SMII

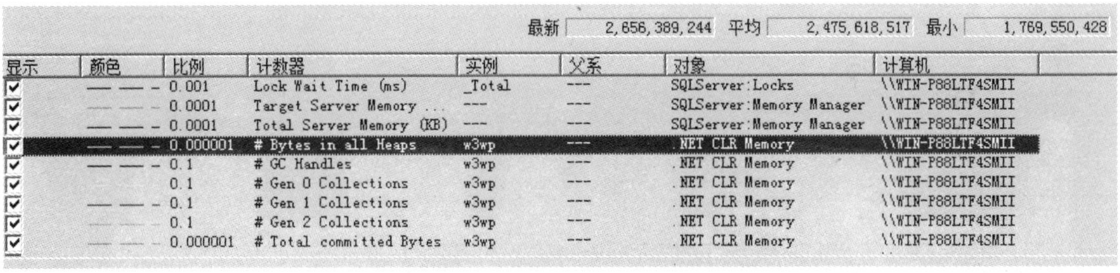

图 5-4　堆内存总大小

.NET 所有堆总内存占用 2.6GB，其中第 0 代堆空间总大小为 55MB，如图 5-5 所示。

最新 55,447,232　平均 61,092,642　最小 20,97

显示	颜色	比例	计数器	实例	父系	对象	计算机
☑		0.000001	# Total reserved Bytes	w3wp	---	.NET CLR Memory	\\WIN-P88LTF4SMII
☑		1.0	% Time in GC	w3wp	---	.NET CLR Memory	\\WIN-P88LTF4SMII
☑		0.0001	Gen 0 heap size	w3wp	---	.NET CLR Memory	\\WIN-P88LTF4SMII
☑		0.0001	Gen 1 heap size	w3wp	---	.NET CLR Memory	\\WIN-P88LTF4SMII
☑		0.000001	Gen 2 heap size	w3wp	---	.NET CLR Memory	\\WIN-P88LTF4SMII
☑		0.000001	Large Object Heap size	w3wp	---	.NET CLR Memory	\\WIN-P88LTF4SMII
☑		1.0	% Privileged Time	w3wp	---	Process	\\WIN-P88LTF4SMII
☑		1.0	% Processor Time	w3wp	---	Process	\\WIN-P88LTF4SMII
☑		1.0	% User Time	w3wp	---	Process	\\WIN-P88LTF4SMII

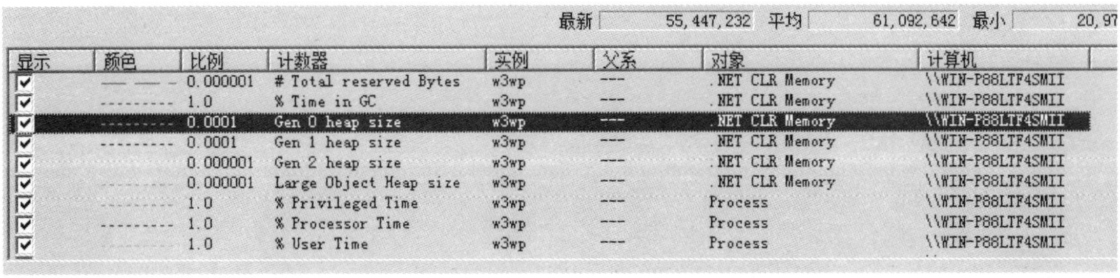

图 5-5　第 0 代堆大小

第 1 代堆空间总大小为 1MB，如图 5-6 所示。

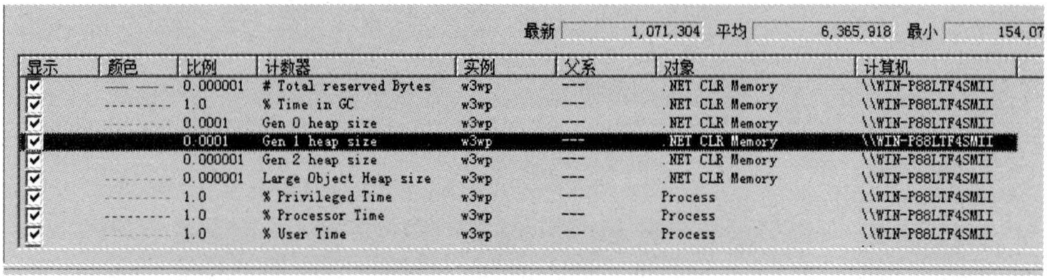

图 5-6　第 1 代堆大小

第 2 代堆空间总大小为 2.4GB，如图 5-7 所示。

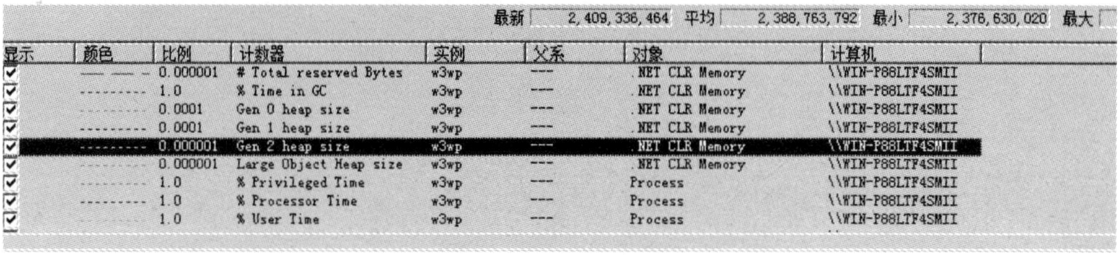

图 5-7　第 2 代堆大小

所有内存都被 2 代堆消耗完了。这个堆表示对象存活时间最长的堆，问题就在于此。这说明内存是被托管内存消耗，而不是被非托管内存用竭了。

大对象堆占用 268MB，如图 5-8 所示。

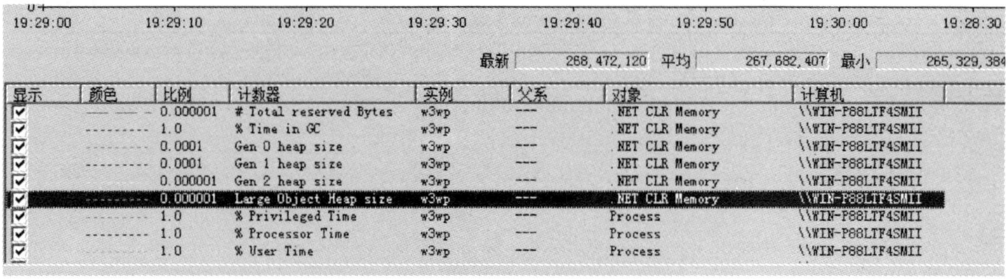

图 5-8　大对象堆大小

看一下 w3wp 内部内存分配情况，w3wp 占用 Private Bytes（独立物理和虚拟内存和）将近 3GB，如图 5-9 所示。

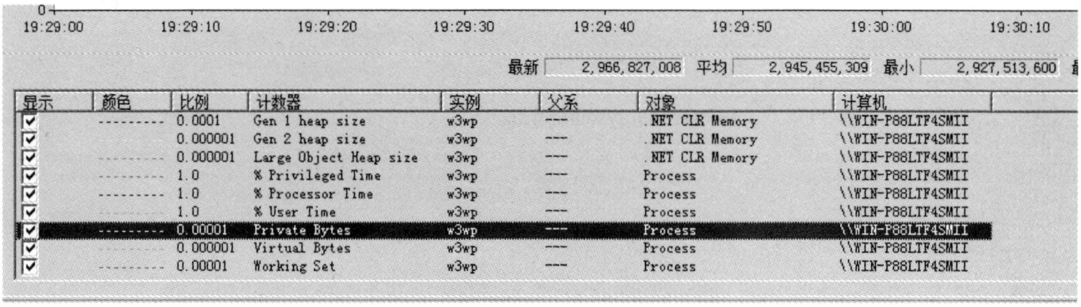

图 5-9　私有内存大小

这个产品从来没有出现过 w3wp.exe 占用内存达到 3GB 的情况，工作集（物理内存）占用 2.9GB 左右，如图 5-10 所示。

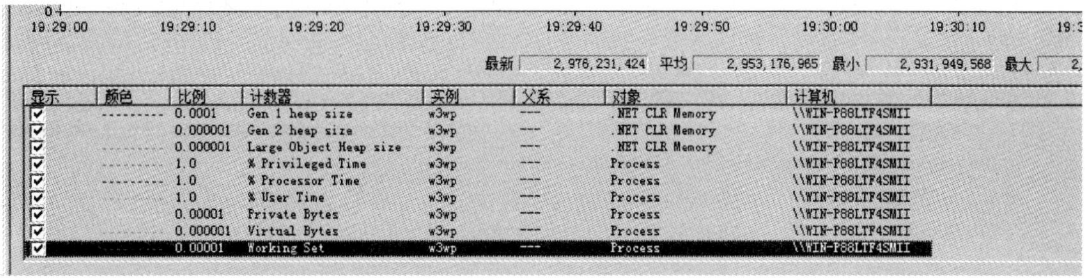

图 5-10　工作集大小

任务管理器中 w3wp 进程占用也是 2.9GB，如图 5-11 所示。

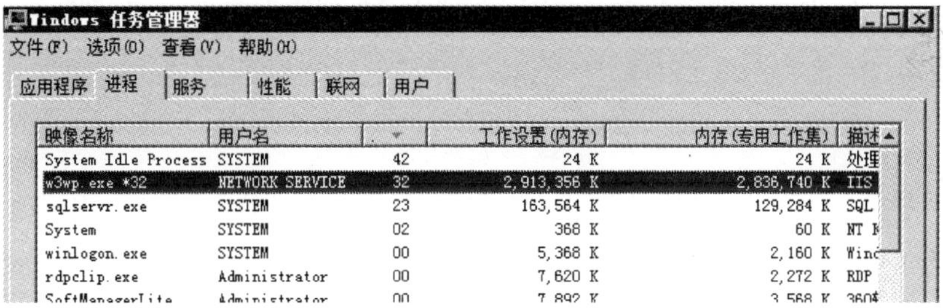

图 5-11　任务管理器显示内存大小

至此问题已经明确，是托管内存使用问题，并且是 w3wp 进程中的第 2 代堆（生存周期长）的对象所致。

一般这种问题主要由两方面导致：

① 使用了 static 或存储到 Cache(Session)中，大量生成对象得不到释放。
② 方法长时间执行，堆栈中的局部大量变量无法释放。

这里是第 2 种。

2．分析定位

对 IIS 主进程 w3wp.exe 抓取了两个 Dump 文件，这两个文件抓取时间间隔为 5min。

第 1 个 Dump 文件 Dump1 中内存情况如下：

```
0:000> !dumpHeap -stat
MT       Count      TotalSize  Class Name
……（省略完事代码）
6edff9ac  2952095    137937248  System.String
26ca09c4  4209417    168376680  King.K.KAP.Report.AggreFunction
6edb6c28  2748694    195505788  System.Object[]
26ca4b84  8644390    207465360  System.Collections.Generic.List`1[[King.K.KAP.Report.ConImplement.Implement, King.K.KAP.Report.ConImplement]]
26ca1730  8644390    414930720  King.K.KAP.Report.ConImplement.Column
```

第 2 个 Dump 文件 Dump2 中内存情况如下：

```
0:000> !dumpHeap -stat
Statistics:
MT          Count       TotalSize  Class Name
……（省略完整代码）
26ca09c4    4130979     165239160  King.K.KAP.Report.AggreFunction
6edb6c28    3163751     223210908  System.Object[]
26ca4b84    14257131    342171144  System.Collections.Generic.List`1[[King.K.KAP.Report.
ConImplement. Implement, King.K.KAP.Report.ConImplement]]
6edff9ac    4367910     558604048  System.String
26ca1730    14247177    683864496  King.K.KAP.Report.ConImplement.Column
```

在 5min 时间内，字符串从 137MB 增加到 558MB，ConImplement.Column 对象集从 414MB 增加到 683MB，共有 14 247 177 个对象。仅 Column 一个对象就占 683MB，比以往的 w3wp.exe 整个进程还大。

以 Column 对象为引导展开分析，如图 5-12 所示是这个对象的对象列表。

```
041367a8  26ca1730      48
041367f0  26ca1730      48
04136838  26ca1730      48
04136880  26ca1730      48
041368c8  26ca1730      48
04136910  26ca1730      48
04136958  26ca1730      48
04136a00  26ca1730      48
04136a90  26ca1730      48
04136af0  26ca1730      48
04136b50  26ca1730      48
04136b98  26ca1730      48
04136c70  26ca1730      48
04136cb8  26ca1730      48
04136d00  26ca1730      48
04136d48  26ca1730      48
04136d90  26ca1730      48
04136e1c  26ca1730      48
04136e64  26ca1730      48
04136eac  26ca1730      48
04136ef4  26ca1730      48
04136f3c  26ca1730      48
04136f84  26ca1730      48
04136fcc  26ca1730      48
04137014  26ca1730      48
0413705c  26ca1730      48
041370a4  26ca1730      48
041370ec  26ca1730      48
```

图 5-12 Column 类型对象列表

图 5-12 中只是部分截图，每一个对象都占 48B，实际有 14 247 177 个这样的对象。随机查看其中一个对象。

```
0:000> !do 04136a90
Name:        King.K.KAP.Report.ConImplement.Column
MethodTable: 26ca1730
EEClass:     26cc0010
Size:        48(0x30) bytes
File:        C:\Windows\Microsoft.NET\Framework\v4.0.30319\Temporary ASP.NET Files\yytpro\
79e71293\f02ca02a\assembly\dl3\d94a4fe6\0091d47e_ba36ce01\King.K.KAP.Report.ConImplement.dll
Fields:
      MT        Field      Offset    Type VT     Attr          Value Name
6ee02978    40000d9       1c     System.Int32  1 instance        13 mIndex
6edff9ac    40000da       4      System.String 0 instance  0af5eab0 mFieldName
6edff9ac    40000db       8      System.String 0 instance  0af5eb58 mDisplayName
```

6edff9ac	40000dc		c	System.String	0 instance 0af5eab0	mReturnFieldName
26ca49f4	40000dd	10	...port.ConImplement]]		0 instance 04136ac0	mExpressList
6edff5e8	40000de	14		System.Object	0 instance 04136ad8	mFieldReturnValue
6ee0662c	40000df	28		System.Boolean	1 instance	1 mDisplay
6ee02978	40000e0	20		System.Int32	1 instance	1 mDataType
6ee0662c	40000e1	29		System.Boolean	1 instance	1 mHideZeroValue
6edff9ac	40000e2	18		System.String	0 instance 0af5ea88	format
6ee02978	40000e3	24		System.Int32	1 instance	1 canSum

发现对象所在的程序集为 King.K.KAP.Report.ConImplement.dll（在这里可以直接通过.net reflector 工具反编译查看源代码）。通过 DebugDialog 分析 dump 文件，看到 54 号线程执行了 7min，如图 5-13 所示。

```
Note - Times include both user mode and kernel mode for each thread
Thread ID: 54   Total CPU Time: 00:07:29.454   Entry Point for Thread: 0x00000000
Thread ID: 61   Total CPU Time: 00:06:40.641   Entry Point for Thread: 0x00000000
Thread ID: 21   Total CPU Time: 00:00:54.911   Entry Point for Thread: 0x00000000
Thread ID: 22   Total CPU Time: 00:00:44.709   Entry Point for Thread: 0x00000000
Thread ID: 23   Total CPU Time: 00:00:42.712   Entry Point for Thread: 0x00000000
```

图 5-13　DebugDialog 报告中监控到的线程运行情况

看一下这个线程的堆栈：

0:054> !clrstack
OS Thread Id: 0xfc0 (54)
Child SP IP   Call Site
2359da94 6ead69c3 [InlinedCallFrame: 2359da94]
2359da90 6eda0580 DomainNeutralILStubClass.IL_STUB_PInvoke(IntPtr, System.String, System. String, Int32, Int32, System.String, Int32, Int32, Int32)
2359da94 6ed17ca7 [InlinedCallFrame: 2359da94] System.Globalization.CompareInfo. Internal CompareString(IntPtr, System.String, System.String, Int32, Int32, System.String, Int32, Int32, Int32)
2359dafc 6ed17ca7 System.Globalization.CompareInfo.Compare(System.String, System.String, System. Globalization.CompareOptions)
2359db18 6f64ccb8 System.String.Equals(System.String, System.StringComparison)
2359db2c 244d5a82 **King.K.KAP.Report.ConImplement.RowObj.get_Item**(System.String)
2359db78 244d57fc **King.K.KAP.Report.ConImplement.Report.AggreDetail**(King.K.KAP. Report. ConImplement.CellRG)
2359dc3c 244d5255 King.K.KAP.Report.ConImplement.Report.AggreProcess()
2359dcd4 244d4945 King.K.KAP.Report.ConImplement.Report.GetBTable(System.Collections. Generic. IList`1<King.K.KAP.Report.DataObject.FilterItemPrint>)
2359dd14 244d48f8 King.K.KAP.Report.ConImplement.Report.GetBTable()
2359dd18 26c6a87f King.K.KAP.ReportModel.SyntaxAnalyse.GetBTable22()
……（省略完整代码）
2359e834 6d7c5a7c [InlinedCallFrame: 2359e834]
2359e8a8 6d7c5a7c DomainNeutralILStubClass.IL_STUB_COMtoCLR(Int32, Int32, IntPtr)
2359ea3c 700925c1 [GCFrame: 2359ea3c]
2359eaac 700925c1 [ContextTransitionFrame: 2359eaac]
2359eae0 700925c1 [GCFrame: 2359eae0]
2359ec38 700925c1 [ComMethodFrame: 2359ec38]

结合前面 54 号线程执行了 7min 的情况，基本上推断是这个线程（堆线代码的代码方法）在执行，并且中间产生了大量的对象所致。下面就针对此堆栈展开分析。

查看堆栈参数：

……（省略完整代码）

2359db2c 244d5a82 King.K.KAP.Report.ConImplement.RowObj.get_Item(System.String)
    PARAMETERS:
        this = <no data>
        strProprty (<CLR reg>) = 0x0af5f0cc

2359db78   244d57fc   King.K.KAP.Report.ConImplement.Report.AggreDetail(King.K.KAP.Report.ConImplement. CellRG)
    PARAMETERS:
        this (0x2359db88) = 0x1103f114
        cg (0x2359db84) = **0x9134ae38**

2359dc3c 244d5255 King.K.KAP.Report.ConImplement.Report.AggreProcess()
    PARAMETERS:
        this (0x2359dc50) = 0x1103f114

……（省略完整代码）

查看 CG 对象的值：

0:000> !do 0x9134ae38
    Name: King.K.KAP.Report.ConImplement.CellRG
    MethodTable: 26ca3fa8
    EEClass: 26cc1d48
    Size: 68(0x44) bytes
    File: C:\Windows\Microsoft.NET\Framework\v4.0.30319\Temporary ASP.NET Files\yytpro\79e71293\f02ca02a\assembly\dl3\d94a4fe6\0091d47e_ba36ce01\King.K.KAP.Report.ConImplement.dll
    Fields:

MT	Field	Offset	Type	VT	Attr	Value	Name
6edff9ac	40000c3	4	System.String	0	instance	00000000	_gExpression
6edff9ac	40000c4	8	System.String	0	instance	0f31e32c	_columnName
26ca44cc	40000c5	c	...port.ConImplement]]	0	instance	9134ae7c	childGroups
26ca0a18	40000c6	10	...port.ConImplement]]	0	instance	9134ae94	**aggres**
6a93a960	40000c7	14	...ow, System.Data]]	0	instance	9134aeac	details
26ca13e4	40000c8	18	...K.GroupColumnInfo	0	instance	00000000	_gColumnInfo
26ca0fa4	40000f8	1c	...rtTableColumnInfo	0	instance	11116080	mRTCI
26ca3bac	40000f9	20	...port.ConImplement]]	0	instance	9134aec4	**mRows**
26ca4394	40000fa	24	...port.ConImplement]]	0	instance	9134aedc	mChildren
6edff5e8	40000fb	28	System.Object	0	instance	0a950260	mGroupValue
26ca3a24	40000fc	2c	...ConImplement.RowObj	0	instance	11f143dc	mRowObjGrp
6ee0662c	40000fd	3c	System.Boolean	1	instance	0	mGroupLeveStop
6ee0662c	40000fe	3d	System.Boolean	1	instance	0	mIsHorizontalLayout
6ee0662c	40000ff	3e	System.Boolean	1	instance	0	mHideGroupSumRow

6edff9ac	4000100	30	System.String	0 instance	0a950260	mHorizontalLayoutKey
6edff9ac	4000101	34	System.String	0 instance	0a950260	mCGKey
6ee02978	4000102	38	System.Int32	1 instance	0	groupLevel

结合 King.K.KAP.Report.ConImplement.dll 反射（Reflector）代码，如图 5-14 所示。

```
protected void AggreDetail(CellReportGroup cg)
{
    foreach (CellReportGroup group in cg.Children)
    {
        this.AggreDetail(group);
    }
    foreach (AggreFunction function in cg.Aggres)
    {
        decimal num = 0M;
        int num2 = 0;
        if ((function.AggregateType == AggregateType.Sum) || (function.AggregateType == AggregateType.NoOut))
        {
            foreach (RowObj obj2 in cg.Rows)
            {
                decimal result = 0M;
                string fieldName = function.FieldName;
                if (obj2[fieldName] == null)
                {
                    fieldName = ReportHelper.GetFieldName(this, fieldName, base.ReportTableColCollection[fieldName].get_Title());
                }
                object fieldReturnValue = obj2[fieldName].FieldReturnValue;
                this.AggreGroup(fieldReturnValue);
                if ((fieldReturnValue != null) && (fieldReturnValue != DBNull.Value))
                {
                    try
                    {
                        if (decimal.TryParse(fieldReturnValue.ToString(), NumberStyles.Float, null, out result))
                        {
                            num2++;
                        }
                    }
                    catch
                    {
                    }
```

图 5-14 反射代码

在代码中，Rows 和 Aggres 控制循环次数属性。还有一个递归调用（第一个 foreach 语句里面调用代码）。

据猜测很可能是循环次数太多，导致此方法执行了 7min，使对象不断增加得不到释放，最终 w3wp 达到近 3GB 大小而崩溃。

接着，取 Rows.Size 和 cg.Aggres.Size 两个属性的值看一下：

0:000> !do 9134ae94
　　　　Name:　　　　　System.Collections.Generic.List`1[[King.K.KAP.Report.AggreFunction, King.K.KAP.Report.ConImplement]]
　　　　MethodTable: 26ca0a18
　　　　EEClass:　　6eb3a530
　　　　Size:　　　　24(0x18) bytes
　　　　File:　　　C:\Windows\Microsoft.Net\assembly\GAC_32\mscorlib\v4.0_4.0.0.0__b77a5c561934e089\mscorlib.dll
　　　　Fields:
　　　　　　　　MT　　　Field　　Offset　　　　　　　Type VT　　　Attr　　　Value Name

6edb6c28	4000bff	4	System.Object[]	0 instance	9134b05c	_items
6ee02978	4000c00	c	System.Int32	1 instance	11	_size
6ee02978	4000c01	10	System.Int32	1 instance	11	_version
6edff5e8	4000c02	8	System.Object	0 instance	00000000	_syncRoot
6edb6c28	4000c03	0	System.Object[]	0 shared	static	_emptyArray

\>\> Domain:Value dynamic statics NYI 00f37f88:NotInit dynamic statics NYI 02690330:NotInit \<\<

外循环 Aggres.Size 有 11 次。

```
0:000> !do 9134aec4
```
Name: System.Collections.Generic.List`1[[King.K.KAP.Report.ConImplement.RowObj, King.K.KAP.Report.ConImplement]]
MethodTable: 26ca3a78
EEClass: 6eb3a530
Size: 24(0x18) bytes
File: C:\Windows\Microsoft.Net\assembly\GAC_32\mscorlib\v4.0_4.0.0.0__b77a5c561934e089\mscorlib.dll
Fields:

MT	Field	Offset	Type VT	Attr	Value	Name
6edb6c28	4000bff	4	System.Object[]	0 instance	172a1398	_items
6ee02978	4000c00	c	System.Int32	1 instance	373164	_size
6ee02978	4000c01	10	System.Int32	1 instance	373164	_version
6edff5e8	4000c02	8	System.Object	0 instance	00000000	_syncRoot
6edb6c28	4000c03	0	System.Object[]	0 shared	static	_emptyArray

\>\> Domain:Value dynamic statics NYI 00f37f88:NotInit dynamic statics NYI 02690330:NotInit \<\<

内循环有 373164 次。

外循环和内循环相乘为 410 万余次。

这就是此方法执行长达 7min 的原因。这里没有静态对象，也没有存储在 Cache 中，主要是由于方法执行时间太长导致的。

5.3.2 Web 服务器运行中突然崩溃原因定位

1．问题描述

系统在使用过程中，发生 IIS 崩溃，所有客户端都无法访问服务器，IIS 不再响应客户端任何请求。但服务器 w3wp 进程仍然在运行，w3wp 进程占用 CPU 资源也恒为 0。

只能重启 IIS 才能使用系统，在重启之前我们做了个 Dump。

2．分析定位

下面就分析一下 Dump 文件的内部情况，查看进程崩溃后所有线程的状态：

```
0:000> !threads
ThreadCount:      47
UnstartedThread:  0
BackgroundThread: 47
PendingThread:    0
DeadThread:       0
```

Hosted Runtime: no

	ID	OSID	ThreadOBJ	State	PreEmptive GC	GC Alloc Context	Domain	Lock Count	APT	Exception
11	1	192c	00e09358	8220	Enabled	00000000:00000000	00e04eb8	0	Ukn	
29	2	16c8	1adb8928	b220	Enabled	00000000:00000000	00e04eb8	0	MTA	(Finalizer)
31	7	1eb4	1adee448	100a220	Enabled	00000000:00000000	00e04eb8	0	MTA	(Threadpool Worker)
32	8	b80	1adfc820	1220	Enabled	00000000:00000000	00e04eb8	0	Ukn	

……（省略完整代码）

63	a	1380	1f474b60	1009220	Enabled	00000000:00000000	00e04eb8	0	MTA	(Threadpool Worker)
64	15	16a4	1f3886b0	1009222	Disabled	41182220:41183fe0	1adfc0c8	2	MTA	(Threadpool Worker) System.StackOverflowException (0c7a0104)
66	23	d50	1f179e00	1009220	Enabled	00000000:00000000	00e04eb8	0	MTA	(Threadpool Worker)
65	24	e5c	1f2ff410	1019220	Enabled	00000000:00000000	00e04eb8	0	Ukn	

……（省略完整代码）

| 71 | 2a | 15f8 | 2b8aaf70 | 1009220 | Enabled | 3d465408:3d466940 | 1adfc0c8 | 1 | MTA | (Threadpool Worker) |
| 72 | 14 | 1f3c | 1f3890c0 | 1019220 | Enabled | 00000000:00000000 | 00e04eb8 | 0 | Ukn | (Threadpool Worker) |

其中有一个异常线程，如下：

```
64   15   16a4 1f3886b0   1009222 Disabled 41182220:41183fe0 1adfc0c8       2 MTA (Threadpool Worker) System.StackOverflowException (0c7a0104)
```

这样的异常主要是执行堆栈溢出错误引发的，通常在存在非常深的递归或无界递归时发生，由死循环或无限递归导致。这个异常可以让 IIS 崩溃，产生异常的线程是 64 号线程。

另一方面，看一下是否有死锁的可能，如下：

```
0:000> !syncblk
Index SyncBlock MonitorHeld Recursion Owning Thread Info    SyncBlock Owner
  343 1ca26654           1         1 1f474150 1860   54     072b19dc ASP.global_asax
  353 1caa2268           1         1 1f436620 1a58   44     091be054 ASP.global_asax
-----------------------------
Total           389
CCW             4
RCW             6
ComClassFactory 0
Free            160
```

在进程崩溃时有两个线程分别占用了两个资源，但没有等待的进程，可以排除线程死锁的可能。

继续关注 System.StackOverflowException 异常，设法从这里找到 IIS 崩溃原因。切换到 64 号线程，看这个线程在做什么：

```
0:000> ~64s
eax=00000000 ebx=312c2a08 ecx=00000000 edx=312c3010 esi=00000002 edi=00000000
eip=76ed014d esp=312c29b8 ebp=312c2a54 iopl=0         nv up ei pl zr na pe nc
cs=0023  ss=002b  ds=002b  es=002b  fs=0053  gs=002b          efl=00000246
ntdll!NtWaitForMultipleObjects+0x15:
76ed014d 83c404          add     esp,4
```

查看其调用堆栈：

```
0:064> !clrstack
Child SP IP      Call Site
312c2b58 76ed014d [FaultingExceptionFrame: 312c2b58]
312c353c 76ed014d [DebuggerU2MCatchHandlerFrame: 312c353c]
312c3508 76ed014d [CustomGCFrame: 312c3508]
312c34dc 76ed014d [GCFrame: 312c34dc]
312c34c0 76ed014d [GCFrame: 312c34c0]
312c36e4 76ed014d [HelperMethodFrame_PROTECTOBJ: 312c36e4] System.RuntimeMethodHandle._InvokeMethodFast(System.IRuntimeMethodInfo, System.Object, System.Object[], System.SignatureStructByRef, System.Reflection.MethodAttributes, System.RuntimeType)
312c3760 6ca6d689 System.RuntimeMethodHandle.InvokeMethodFast(System.IRuntimeMethodInfo, System.Object, System.Object[], System.Signature, System.Reflection.MethodAttributes, System.RuntimeType)
312c37b4 6ca6d3d0 System.Reflection.RuntimeMethodInfo.Invoke(System.Object, System.Reflection.BindingFlags, System.Reflection.Binder, System.Object[], System.Globalization.CultureInfo, Boolean)
312c37f0 6ca6bfed System.Reflection.RuntimeMethodInfo.Invoke(System.Object, System.Reflection.BindingFlags, System.Reflection.Binder, System.Object[], System.Globalization.CultureInfo)
312c3814 6ca463f8 System.Reflection.RuntimePropertyInfo.GetValue(System.Object, System.Reflection.BindingFlags, System.Reflection.Binder, System.Object[], System.Globalization.CultureInfo)
312c3838 6ca463ac System.Reflection.RuntimePropertyInfo.GetValue(System.Object, System.Object[])
312c3844 233897d5 AjaxPro.DataObjectCollectionConverter.Serialize(System.Object, System.Text.StringBuilder)
312c38c0 23388e51 AjaxPro.DataObjectConverter.Serialize(System.Object, System.Text.StringBuilder)
312c3904 22e1736d AjaxPro.IJavaScriptConverter.TrySerializeValue(System.Object, System.Type, System.Text.StringBuilder)
312c3924 1f7a33c5 AjaxPro.JavaScriptSerializer.Serialize(System.Object, System.Text.StringBuilder)
312c3980 233898b8 AjaxPro.DataObjectCollectionConverter.Serialize(System.Object, System.Text.StringBuilder)
312c39fc 23388e51 AjaxPro.DataObjectConverter.Serialize(System.Object, System.Text.StringBuilder)
312c3a40 22e1736d AjaxPro.IJavaScriptConverter.TrySerializeValue(System.Object, System.Type, System.Text.StringBuilder)
312c3a60 1f7a33c5 AjaxPro.JavaScriptSerializer.Serialize(System.Object, System.Text.StringBuilder)
312c3abc 233898b8 AjaxPro.DataObjectCollectionConverter.Serialize(System.Object, System.Text.StringBuilder)
312c3b38 23388e51 AjaxPro.DataObjectConverter.Serialize(System.Object, System.Text.StringBuilder)
312c3b7c 22e1736d AjaxPro.IJavaScriptConverter.TrySerializeValue(System.Object, System.Type, System.Text.StringBuilder)
312c3b9c 1f7a33c5 AjaxPro.JavaScriptSerializer.Serialize(System.Object, System.Text.StringBuilder)
312c3bf8 233898b8 AjaxPro.DataObjectCollectionConverter.Serialize(System.Object, System.Text.
```

StringBuilder)

……（省略了 70 页的堆栈，大概调用了近 1000 次）
　　312fd198 23388e51 AjaxPro.DataObjectConverter.Serialize(System.Object, System.Text.StringBuilder)
　　312fd1dc 22e1736d AjaxPro.IJavaScriptConverter.TrySerializeValue(System.Object, System.Type, System.Text.StringBuilder)
　　312fd1fc 1f7a33c5 AjaxPro.JavaScriptSerializer.Serialize(System.Object, System.Text.StringBuilder)
　　312fd258 233898b8 AjaxPro.DataObjectCollectionConverter.Serialize(System.Object, System.Text.StringBuilder)
　　312fd2d4 23388e51 AjaxPro.DataObjectConverter.Serialize(System.Object, System.Text.StringBuilder)
　　312fd318 22e1736d AjaxPro.IJavaScriptConverter.TrySerializeValue(System.Object, System.Type, System.Text.StringBuilder)
　　312fd338 1f7a33c5 AjaxPro.JavaScriptSerializer.Serialize(System.Object, System.Text.StringBuilder)
　　312fd394 1f7a320a AjaxPro.JavaScriptSerializer.Serialize(System.Object)
　　312fd42c 1f7ab1e8 King.K.KAP.Aop.Dynamic.LocalCallDynamicProxyImpl.Invoke(System.Runtime. Remoting. Messaging.IMessage)
　　312fd490 6ca1a25e System.Runtime.Remoting.Proxies.RealProxy.PrivateInvoke(System.Runtime. Remoting.Proxies.MessageData ByRef, Int32)
　　312fd720 6d582356 [TPMethodFrame: 312fd720] King.K.AM.Interface.IHandleService. ExternalValidate (King. K.AM.DataObject.HandleDataObject)
　　……（省略完整代码）
　　312fe508 6ca6d689 System.RuntimeMethodHandle.InvokeMethodFast(System.IRuntimeMethodInfo, System.Object, System.Object[], System.Signature, System.Reflection.MethodAttributes, System.RuntimeType)
　　312fe55c 6ca6d37c System.Reflection.RuntimeMethodInfo.Invoke(System.Object, System.Reflection. BindingFlags, System.Reflection.Binder, System.Object[], System.Globalization.CultureInfo, Boolean)
　　312fe598 6ca6bfed System.Reflection.RuntimeMethodInfo.Invoke(System.Object, System.Reflection. BindingFlags, System.Reflection.Binder, System.Object[], System.Globalization.CultureInfo)
　　312fe5bc 6ca73284 System.Reflection.MethodBase.Invoke(System.Object, System.Object[])
　　312fe5c8 22e1413c AjaxPro.AjaxProcHelper.Run()
　　312fe728 22e136cc AjaxPro.AjaxSyncHttpHandler.ProcessRequest(System.Web.HttpContext)
　　312fe73c 69fb1b75 System.Web.HttpApplication+CallHandlerExecutionStep.System.Web. HttpApplication. IExecutionStep.Execute()
　　312fe768 69fc611c System.Web.HttpApplication.ExecuteStep(IExecutionStep, Boolean ByRef)
　　312fe7ac 69fdc10e System.Web.HttpApplication+ApplicationStepManager.ResumeSteps (System. Exception)
　　312fe800 69f92cdd System.Web.HttpApplication.System.Web.IHttpAsyncHandler.BeginProcessRequest (System.Web.HttpContext, System.AsyncCallback, System.Object)
　　312fe81c 69fda8f2 System.Web.HttpRuntime.ProcessRequestInternal(System.Web.HttpWorkerRequest)
　　312fe850 69fda63d System.Web.HttpRuntime.ProcessRequestNoDemand(System.Web. HttpWorkerRequest)
　　312fe860 69fd9c3d System.Web.Hosting.ISAPIRuntime.ProcessRequest(IntPtr, Int32)
　　312fe864 6a5f5a7c [InlinedCallFrame: 312fe864]
　　312fe8d8 6a5f5a7c DomainNeutralILStubClass.IL_STUB_COMtoCLR(Int32, Int32, IntPtr)
　　312fea6c 6d5825c1 [GCFrame: 312fea6c]
　　312feadc 6d5825c1 [ContextTransitionFrame: 312feadc]
　　312feb10 6d5825c1 [GCFrame: 312feb10]
　　312fec68 6d5825c1 [ComMethodFrame: 312fec68]

可以看到下面这几个方法产生了近1000次的递归调用，最后崩溃：

  ……（省略完整代码）
  312fd258 233898b8 AjaxPro.DataObjectCollectionConverter.Serialize(System.Object, System.Text.StringBuilder)
  312fd2d4 23388e51 AjaxPro.DataObjectConverter.Serialize(System.Object, System.Text.StringBuilder)
  312fd318 22e1736d AjaxPro.IJavaScriptConverter.TrySerializeValue(System.Object, System.Type, System.Text.StringBuilder)
  312fd338 1f7a33c5 AjaxPro.JavaScriptSerializer.Serialize(System.Object, System.Text.StringBuilder)
  ……（省略完整代码）

根据整个堆栈调用顺序，找到开发负责人，很快定位到原因：

```
//以下为伪代码
Public override HandleVObject(string id, string type)
{
    String [] strArgs=new string[1]
    stringArgs[0]=type;
    int num=0;
    HandleVObject obj=Service.GetInstance(strArgs);
    String[] sourceIdArgs=this.Session["args"]
    foreach(string id in sourceIdArgs)
    {
        HandleVODetailObject detail=new HandleVODetailObject();
        detail.ID=Guid.NewGuid();
        detail.Code=CODE.NewCode();
        detail.Handle=obj;
        obj.Details.Add(obj);
    }
    return obj;
}
```

上面的伪代码中，主类添加了很多子类作为明细记录，每个子类（detail）又引用了父类，造成循环引用，导致序列化程序死循环。

一旦线程陷入死循环，会导致IIS堆栈溢出错误，通常在存在非常深的递归或无界递归时发生，由死循环或无限递归导致，这个异常可以让IIS崩溃。错误会导致单个请求失败，继而引起整个服务器崩溃，要避免这种情况。崩溃后进程自动重启

这个堆栈溢出还是相对容易解决的，因为w3wp进程没有崩溃或自动重启。在这之后遇到的另一个客户案例（分析思路与上述案例基本一样，就不再单独作为案例讲解了），除了报堆栈溢出，崩溃后还自动重启并创建了一个新的w3wp进程，也就是说崩溃后会马上破坏现场。尝试用WinDbg或DebugDialog的Crash自动捕捉也捕捉不到，最后通过如下方式解决：首先用Visual Studio断点跟踪到刚刚执行完堆栈异常的地方（一定要排除异常后），执行完断点代码后保持调试状态（不要按[F5][F10][F11]）继续执行，然后通过DebugDialog进行Dump文件（这时由于Visual Studio附加了w3wp进程，还无法使用WinDbg直接附加进程调试），得到Dump文件再用WinDbg进行分析。

5.3.3 DevGrid 控件 EventHandler 事件泄漏内存

5.3.3.1 问题描述

这是人力资源系统（C/S）在运行时出现的一个严重内存泄漏问题：系统菜单是以页签的形式展现，每打开一次员工档案，应用程序主进程内存（通过任务管理器查看）就增加 30MB，连续打开/关闭 10 次，则进程内存会立即增加 300MB，且系统所有操作都运行缓慢。

5.3.3.2 分析定位

使用 WinDbg 工具分析应用程序进程，定位原因。查看应用程序主进程 HRMain.exe 中的对象情况，如下：

```
0:020>!dumpHeap -stat
……（省略完整代码）
0b1c5f14        8        2080 HRRY.HRPersonMain
0b065db4        8       82200 DevExpress.XtraGrid.Views.Grid.GridView
79329d90     6721     2253472 System.EventHandler
……（省略完整代码）
```

查看其中一个 GridView 在内存中的引用关系，如下：

```
0:020> !gcroot 0d6e8d28
Scan Thread 0 OSTHread 29c
Scan Thread 3 OSTHread 46c
Scan Thread 8 OSTHread 10a0
Scan Thread 6 OSTHread fd0
Scan Thread 9 OSTHread 714
Scan Thread 11 OSTHread 4e8
Scan Thread 12 OSTHread 224
Scan Thread 13 OSTHread 904
Scan Thread 14 OSTHread b60
Scan Thread 18 OSTHread 6d8
DOMAIN(001732E0):HANDLE(Pinned):4413a8:Root:0250a028(System.Object[])->
0da4822c(System.EventHandler)->
0da4801c(System.Object[])->
0d6e90bc(System.EventHandler)->
0d6e8d28(DevExpress.XtraGrid.Views.Grid.GridView)
```

查看 System.EventHandler 所指向的目标（一般为一个方法）：

```
0:020> !do 0d6e90bc
Name: System.EventHandler
MethodTable: 79329d90
EEClass: 790c39d0
Size: 32(0x20) bytes
 (C:\WINDOWS\assembly\GAC_32\mscorlib\2.0.0.0__b77a5c561934e089\mscorlib.dll)
Fields:
      MT    Field   Offset                 Type VT     Attr    Value Name
```

79330740	40000ff	4	System.Object	0 instance	0d6e8d28	_target
7932ff98	4000100	8	...ection.MethodBase	0 instance	00000000	_methodBase
793333ec	4000101	c	System.IntPtr	1 instance	**a4c9770**	**_methodPtr**
793333ec	4000102	10	System.IntPtr	1 instance	0	_methodPtrAux
79330740	400010c	14	System.Object	0 instance	00000000	_invocationList
793333ec	400010d	18	System.IntPtr	1 instance	0	_invocationCount

查看这个方法句柄指向的方法名：

0:020> !ip2md **a4c9770**
Failed to request MethodData, not in JIT code range

[

0:020> dd a4c9770
0a4c9770 064924b8 45e8900b e96f9a97 f5ee874c
0a4c9780 eb00b000 eb03b070 eb05b06c eb08b068
0a4c9790 eb0bb064 eb0eb060 eb11b05c eb14b058
0a4c97a0 eb17b054 eb1ab050 eb1db04c eb20b048
0a4c97b0 eb23b044 eb26b040 eb29b03c eb2cb038
0a4c97c0 eb2fb034 eb32b030 eb35b02c eb38b028
0a4c97d0 eb3bb024 eb3eb020 eb41b01c eb44b018
0a4c97e0 eb47b014 eb4ab010 eb4db00c eb50b008
0:020> !dumpmd e96f9a97

e96f9a97 is not a MethodDesc

]

这条路走不通，继续尝试其他的办法。查看方法的内部代码，如下：

0:020> !U a4c9770
Unmanaged code
0a4c9770 b82449060b   mov   eax,offset <Unloaded_ure.dll>+0xb064923 (**0b064924**)
0a4c9775 90   nop
0a4c9776 e845979a6f   call   mscorwks!PrecodeRemotingThunk (79e72ec0)
0a4c977b e94c87eef5   jmp   <Unloaded_ure.dll>+0x3b1ecb (003b1ecc)
0a4c9780 00b000eb70b0   add   byte ptr [eax-4F8F1500h],dh
0a4c9786 03eb   add   ebp,ebx
0a4c9788 6c   ins   byte ptr es:[edi],dx
0a4c9789 b005   mov   al,5
0a4c978b eb68   jmp   <Unloaded_ure.dll>+0xa4c97f4 (0a4c97f5)
0a4c978d b008   mov   al,8

在上面的反汇编代码中，很多情况下第一句的地址就是方法名。
看一下以下地址的方法名：

0:020> !dumpmd 0b064924
Method   Name: DevExpress.XtraGrid.Views.Grid.GridView.**OnLocalizer_Changed**(System.Object, System.EventArgs)

Class: 0af4ab48
MethodTable: **0b065db4**
mdToken: 06000781
Module: 091407a4
IsJitted: no
CodeAddr: ffffffff

不出所料，就是 OnLocalizer_Changed 事件导致了整个窗体泄漏。

又查了几个类似的 EventHandler，跟踪结果与上面一致，也是定位到 OnLocalizer_ Changed 事件。

进一步验证方法名：

0:020> !dumpmt -md 0b065db4
EEClass: 0af4ab48
Module: 091407a4
Name: DevExpress.XtraGrid.Views.Grid.GridView
mdToken: 0200003a (C:\WINDOWS\assembly\GAC_MSIL\DevExpress.XtraGrid.v9.2\9.2.6.0__b88d1754d700e49a\DevExpress.XtraGrid.v9.2.dll)
BaseSize: 0x258
ComponentSize: 0x0
Number of IFaces in IFaceMap: 14
Slots in VTable: 1124

MethodDesc Table
　　Entry MethodDesc   JIT Name
    7a546998   7a460d3c   PreJIT System.ComponentModel.Component.ToString()
    79286ac0   7910494c   PreJIT System.Object.Equals(System.Object)
    79286b30   7910497c   PreJIT System.Object.GetHashCode()
    ……（省略完整代码）
    0a4c339d   0b064d00   NONE DevExpress.XtraGrid.Views.Grid.GridView.SetScrollingState()
    0b418b38   0b064358   JIT DevExpress.XtraGrid.Views.Grid.GridView.SetDefaultState()
    0a4b55a8   09c9a344   JIT DevExpress.XtraGrid.Views.Base.BaseView.OnEndInit()
    0b416378   0b064340   JIT DevExpress.XtraGrid.Views.Grid.GridView.CheckInfo()
    0b3433f0   0b0640e8   JIT DevExpress.XtraGrid.Views.Grid.GridView.OnGridControlChanged(DevExpress.XtraGrid.GridControl)
    0a4c357d   **0b064924**   NONE DevExpress.XtraGrid.Views.Grid.GridView.**OnLocalizer_Changed**(System.Object, System.EventArgs)
    0b419978   0b064ddc   JIT DevExpress.XtraGrid.Views.Grid.GridView.SetDataSource (System.Windows.Forms.BindingContext, System.Object, System.String)
    0b343488   0b063e6c   JIT DevExpress.XtraGrid.Views.Grid.GridView.CreateNullViewInfo()
    0af3d268   09c9a374   JIT DevExpress.XtraGrid.Views.Base.BaseView.SetupInfo()
    0b346718   09c9a37c   JIT DevExpress.XtraGrid.Views.Base.BaseView.OnFormatConditionChanged (System.Object, System.ComponentModel.CollectionChangeEventArgs)
    ……（省略完整代码）

定位到 OnLocalizer_Changed 事件后，通过 Reflector 查看一下 GridLocalizer 的源代码，

如下：

```
[ToolboxItem(false)]
public class GridLocalizer
{
    // Fields
    private static GridLocalizer active;
    private static EventHandler ActiveChanged;

    // Events
    public static    event EventHandler ActiveChanged;

    // Methods
    static GridLocalizer();
    public GridLocalizer();
    public static GridLocalizer CreateDefaultLocalizer();
    public virtual string GetLocalizedString(GridStringId id);
    public static void RaiseActiveChanged();

    // Properties
    [Browsable(false)]
    public static GridLocalizer Active { get; set; }
    public virtual string Language { get; }
}
```

以上即为整个跟踪过程。

这个对象中有个 static 的 EventHandler，一旦 ActiveChanged 事件被注册，GridLocalizer 类不释放会导致所有的 GridView 不释放，更进一步则会导致引用它的 Form 和 Form 下的所有内容都不释放。在这种情况下，即使关闭页签也不会释放，所以重复打开/关闭只会增加内存。

根据上述的这些跟踪，在 DevGrid 官方网站上搜索后发现，正好此问题已经出了补丁，专门用于解决 ActiveChanged 内存泄漏问题，最终通过升级 DevGrid 程序集版本解决了案例中的问题。

正常情况下的一般解决方案如下：

ActiveChanged 注册（+=）是没有问题的，在页签关闭时，窗体卸载（Unload）事件中，要取消注册（-=）ActiveChanged 事件。

这个案例说明，在遇到问题时并不一定都是开发人员的责任，一些第三方组件甚至是系统组件也可能会有代码隐患。控件提供商出补丁是否及时，是确保产品正常使用的关键。

5.3.4 Session 陷阱及正确使用

1．问题描述

本案例也是一个内存问题。在 32 位系统下，Web 服务器进程占用内存达到 1GB（平常都在 600MB 以内）。

一般情况在 32 位系统下，w3wp 达到 1.2GB 或 64 位系统下达到 2.5GB 就基本接近峰值，可能会导致 w3wp 进程不稳定。

2. 分析定位

对服务器进程做了 dump 文件，并进行内存分析。先看一下堆对象统计：

```
0:037> !dumpHeap -stat
------------------------------
Heap 0
total 2459572 objects
------------------------------
Heap 1
total 2553361 objects
------------------------------
Heap 2
total 2427413 objects
------------------------------
Heap 3
total 2552206 objects
------------------------------
total 9992552 objects
Statistics:
      MT      Count      TotalSize Class Name
... ... ... ...
715e5b90    239836      12471472 System.Collections.Generic.Dictionary`2[[System.String, mscorlib],[System.Object, mscorlib]]
715e5d68     64723      13288596 System.Collections.Generic.Dictionary`2+Entry[[System.String, mscorlib], [System.Object, mscorlib]][]
715ea8f0    731025      17544600 System.Collections.ArrayList
0203481c     41419      19549768 King.K.AA.DataObject.InventoryDataObject
71bfc7e8      1939      23086100 System.Decimal[]
715c4eec    376496      52588544 System.Object[]
017a03d8       335     115001600 Free
715e88c0   6203878     353412904 System.String
Total 9992552 objects
```

字符串占用了 353M 空间，比一般情况下稍大。

大多时候整个 w3wp.exe 进程为 300M 左右，而这里仅 w3wp 中的字符串对象就为 353M。

再使用 dumpheap –mt 715e88c0，查看所有这些字符串对象占用的空间大小。挑选几个 String 对象，使用 gcroot 命令查看一下对象的引用关系：

```
Scan Thread 37 OSTHread 4e0
ESP:23d6e6c8:Root:02bb081c(System.Threading.Thread)->
10c0b338(System.Threading.ExecutionContext)->
10c0b35c(System.Runtime.Remoting.Messaging.IllogicalCallContext)->
10c0afd8(System.Web.HttpContext)->
10c0b598(System.Web.RequestTimeoutManager+RequestTimeoutEntry)->
32383e90(System.Web.RequestTimeoutManager+RequestTimeoutEntry)->
```

```
32383b48(System.Web.HttpContext)->
32382ec0(System.Web.Hosting.ISAPIWorkerRequestInProcForIIS7)->
0e29c310(System.Web.HttpWorkerRequest+EndOfSendNotification)->
0e29bfdc(System.Web.HttpRuntime)->
0e29c090(System.Web.RequestTimeoutManager)->
0e29c0b4(System.Object[])->
0e29c120(System.Web.Util.DoubleLinkList)->
25d3dfc8(System.Web.RequestTimeoutManager+RequestTimeoutEntry)->
25d3da08(System.Web.HttpContext)->
25d4554c(System.Collections.Hashtable)->
25d45584(System.Collections.Hashtable+bucket[])->
25d45780(System.Web.SessionState.HttpSessionState)->
25d45758(System.Web.SessionState.HttpSessionStateContainer)->
0ecd886c(System.Web.SessionState.SessionStateItemCollection)->
0ecd88cc(System.Collections.Hashtable)->
02c0aa8c(System.Collections.Hashtable+bucket[])->
06c45c84(System.Collections.Specialized.NameObjectCollectionBase+NameObjectEntry)->
06c45c50(System.Collections.Generic.Dictionary`2[[System.String, mscorlib],[System.Object, mscorlib]])->
2e5863f0(System.Collections.Generic.Dictionary`2+Entry[[System.String, mscorlib], [System.Object, mscorlib]][])->
0ef02118(System.Collections.Generic.List`1[[King.K.GL.DataObject.TchangeProssesForm.TchangeProssesFormDObject, King.K.GL.DataObject]])->
123afae0(System.Object[])->
07c31428(King.K.GL.DataObject.TchangeProssesForm.TchangeProssesFormDObject)->
07c3158c(System.Collections.ArrayList)->
07c318dc(System.Object[])->
07c31994(System.String)
```

可见，在 Session 中存储了一个 HashTable（内含 Dicionry）结构，然后 HashTable 中又引用了一个 TchangeProssesFormDObject 列表。

统计了一下，这里的 TchangeProssesFormDObject 有 37000 多个，占用内存为 219MB 左右。

```
0:037> !objsize 123afae0
sizeof(123afae0) =    219303756 (   0xd124f4c) bytes (System.Object[])

0:037> !objsize 0ecd88cc
sizeof(0ecd88cc) =    219961208 (   0xd1c5778) bytes (System.Collections.Hashtable)
```

根据对象引用关系及代码堆栈找到开发人员代码，如下：

```
protected override void InitControl()
{
    try
    {
        ……（省略完整代码）
        if (!IsPostBack)
        {
```

```
……（省略完整代码）
                    Dictionary<string, object> dic = Session["TchangeProssesformDO"] as
Dictionary<string, object>;

                    list = tchangeProcessformService.GenerateTrialInfo((TchangeProssesFormData
Object)dic["step1"], (List<TransformCDO>)dic["step2"]);
                    ……（省略完整代码）
                    dic.Add("step3", list);
                    string sessionName = Guid.NewGuid().ToString();
                    Session[sessionName] = dic;    //将 DataTable 放入 Session 中

                    return sessionName;
                }
                ……（省略完整代码）
            }
```

上面的代码明显存在很多问题，Session 中不建议存储大量数据，且 Session 的存储键 Key 最好不要用 Guid。

总结一下 Session 使用建议：

◆ 不要用随机 Guid 值作为 Session 的 Key。

使用 Guid.NewGuid 方法作为 Session 的 Key，会导致当前用户每次访问都生成一个对象放到 Session 中，放到 Session 中的对象永远不会释放，直到当前用户注销。正确做法是：同一个用户要永远使用同一个 Key，这样可以保证当前用户多次往 Session 中写值时可以覆盖前面的值，或判断是否已经存在相同数据。

◆ 增加 Session 清理机制。

当用户退出此页面时，要清除当前用户 Session 中的内容。增加 Session 数据清理机制，可以避免这 219MB 内存的占用。

◆ 特殊大数据要用 Cache 对象代替 Session。

如果有多个用户共享，则建议用 Cache 对象存储。Session 是为每个用户单独存储一份，如果 10 个用户同时登录每人都占用 219MB，那么 w3wp 中光这一个对象就达到 1GB 多，这样会导致 w3wp 内存过大崩溃（或自动回收）。

5.3.5　WinDbg 内存泄漏+异常检测案例

1．问题描述

客户在使用过程中，发现内存一直在减少，使用一段时间之后，可用内存从 2GB 减少到不足 1GB，其他都正常。

2．分析定位

这次的 Dump 文件只有 700MB，但其中托管堆占用了 366MB。在托管堆占用的 366MB 中，占用内存资源最多的一个 String 对象占用了多达 225MB。这个是由于异常引起的内存问题，具体分析过程如下。

客户使用期间，我对服务器做了一些监控跟踪，跟踪结果如下：

CPU：CPU 占用时间平均为 29%，其中用户时间 28%，内核时间 1%。可见 CPU 不存在问题。

磁盘：%Disk Time 平均为 5%，平均等待队列 0.11（不到 1 个）。磁盘也不存在问题。

说明：如果做了磁盘阵列，则%DiskTime 计数器不准确，只看磁盘队列即可。

内存：Page Faults/s 不到 1000/s，Page Reads/s 平均 3 次/s，Pages/s 平均 16/s，都在正常范围之内。需要注意的是，可用内存从 2.1GB 变为 1.4GB，呈平均减少趋势，可能是正常缓存使用导致，也可能是内存泄漏。

对内存方面进行跟踪，以下是采集的内存占用情况：

.NET 托管堆占用呈增加趋势，w3wp 进程 private bytes 与.NET 托管堆增长规律一致，说明是托管内存问题。操作系统的可用内存占用呈稳定减少趋势，如图 5-15 所示。

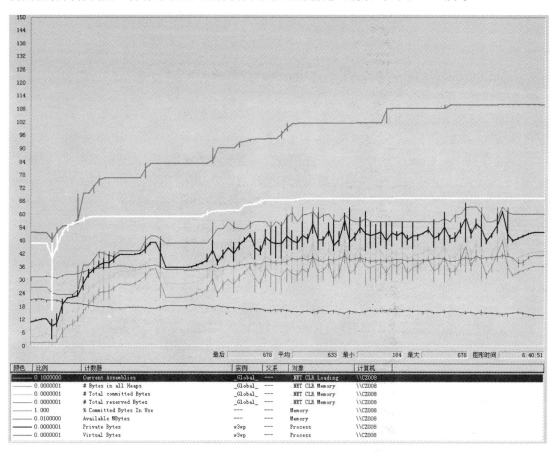

图 5-15 内存计数器

对进程内部内存对象资源占用情况进行一下分析，查看 w3wp.exe 进程情况和.NET 托管堆大小：

```
0:000> !eeheap -gc
Number of GC Heaps: 2
------------------------------
Heap 0 (000e8990)
```

generation 0 starts at 0x361b8794
generation 1 starts at 0x36195cb4
generation 2 starts at 0x02910038
ephemeral segment allocation context: none
　segment　　begin allocated　　size
02910000 02910038　0690fe38 0x03fffe00(67108352)
1a530000 1a530038　1e52ff14 0x03fffedc(67108572)
36060000 36060038　36436c8c 0x003d6c54(4025428)
Large object Heap starts at 0x0a910038
　segment　　begin allocated　　size
0a910000 0a910038　0c3f8368 0x01ae8330(28214064)
24e30000 24e30038　266c02e8 0x018902b0(25756336)
Heap Size　0xb74ef10(192212752)

Heap 1 (000e99b0)
generation 0 starts at 0x2b047608
generation 1 starts at 0x2b010038
generation 2 starts at 0x06910038
ephemeral segment allocation context: none
　segment　　begin allocated　　size
06910000 06910038　0a90456c 0x03ff4534(67061044)
1e610000 1e610038　21da5db0 0x03795d78(58285432)
2b010000 2b010038　2b2c9840 0x002b9808(2856968)
Large object heap starts at 0x0c910038
　segment　　begin allocated　　size
0c910000 0c910038　0d67d840 0x00d6d808(14080008)
22e30000 22e30038　23dd5e88 0x00fa5e50(16408144)
33e30000 33e30038　34cc1fb0 0x00e91f78(15277944)
Heap Size　0xa5e9084(173969540)

GC Heap Size　0x15d37f94(**366182292**)

托管堆占用总大小约 366MB。

查看堆中对象：

```
0:000> !dumpheap -stat
MT       Count    TotalSize Class Name
7ae3ac30     1           12 System.Drawing.ColorConverter
7a5e7038     1           12 System.Diagnostics.PerformanceCounterCategoryType
1288579c  2368       511488 King.KAP.SystemObject.TableColumnInfo
0fc19a80  3366       538560 King.K.SM.Setting.DO.TDO
58560 King.KAP.Domains.DataObjectInfo
119851c4  5606       672720 King.K.AA.DataObject.UDO
7930454c 15556       933360 System.Reflection.ParameterInfo
79332cc0 11729      1003160 System.Int32[]
79333594  1135      1210668 System.Byte[]
```

……（省略完整代码）

```
1044ab9c      34326       1373040 King.KAP.Domains.EnumItem
  79315bf8     5215       1550692 System.Collections.Generic.Dictionary`2+Entry[[System.String,
mscorlib],[System.Object, mscorlib]][]
  10be5f90     3279       1639500 King.K.PU.DataObject.PIDO
  7932fde0    30029       1681624 System.Reflection.RuntimeMethodInfo
  793159e8    41432       2154464 System.Collections.Generic.Dictionary`2[[System.String, mscorlib],
[System.Object, mscorlib]]
  10be7010     3001       2844948 King.K.PU.DO.PIDetailDO
  79332b54   134893       3237432 System.Collections.ArrayList
  1288e160    65628       3937680 King.KAP.Domains.DObjectInfo
  79333274     6045       5267784 System.Collections.Hashtable+bucket[]
  793042f4    92960       7935448 System.Object[]
  000e82c0      304      86426596      Free
  79330b24  1246264     225210272 System.String
Total 2140685 objects
```

在托管堆占用的366MB中,占用字节排名第二位的System.Object[]仅占不到8MB,而排名第一位的String对象占用了多达225MB,且总个数为1246264个。

注意:往往产生问题的并不是字符串对象本身,但可以把字符串作为突破口。导致内存问题的也往往不是一个对象类型,由于它们是引用的关系,可能是一系列对象都有关联,比如ArrayList对象中包含了Object,而Object对象中又包含了String对象,则这三个对象可能都有涉及。

查看字符串方法表:

```
0:000> !dumpheap -mt 79330b24 -min 50000
------------------------------
Heap 0
   Address       MT        Size
030af974 79330b24       50340
030bc360 79330b24       51728
030c8d70 79330b24       52656
030d6bb0 79330b24       53580
030e52b4 79330b24       55428
……(省略完整代码)
030f2dd0 79330b24       55888
03eb2f2c 79330b24       70764
03ec4a88 79330b24       71232
03ed7390 79330b24       71696
055ae870 79330b24       50980
1b9f10a8 79330b24       53160
……(省略完整代码)
------------------------------
Heap 1
   Address       MT        Size
07212e94 79330b24       68452
```

```
072328e4 79330b24      50804
0723ff70 79330b24      51268
0724ebe4 79330b24      52192
……（省略完整代码）
0725d928 79330b24      53120
0726aef8 79330b24      54040
07279328 79330b24      54504
07287980 79330b24      54964
072985dc 79330b24      56352
1eda3b4c 79330b24      51692
1ff141f0 79330b24      50308
203995dc 79330b24      75520
……（省略完整代码）                    //有很多对象，省略大部分。
------------------------------
```

查看其中一个对象内容：

```
0:000> !do 03eb2f2c
Name: System.String
MethodTable: 79330b24
EEClass: 790ed65c
Size: 70764(0x1146c) bytes
  (C:\WINDOWS\assembly\GAC_32\mscorlib\2.0.0.0__b77a5c561934e089\mscorlib.dll)
String: 2013-10-30 10:46:05 [记载] TID:13  delete    from Account

 insert into Account_T(id,
ID,Accountcode,Accountname,
depth,createTime) values ('01a73657-5d31-4366-b91d-3aacc4326b7e',
'01a73657-5d31-4366-b91d-3aacc4326b7e','1001',?????,
'1','2013-10-30 10:46:05');

 insert into Account_T(id,
ID,Accountcode,Accountname,
depth,createTime) values ('03ab5dbd-7f35-4831-a1e0-92e96da23be3',
'03ab5dbd-7f35-4831-a1e0-92e96da23be3','3201',?????,
'1','2013-10-30 10:46:05');

……（省略完整代码）

 insert into Account_T(id,
ID,Accountcode,Accountname,
depth,createTime) values ('076e19fb-0c7d-4d45-9808-454b364478e0',
'076e19fb-0c7d-4d45-9808-454b364478e0','2501',?????,
'1','2013-10-30 10:46:05');

 insert into Account_T(id,
ID,Accountcode,Accountname,
```

depth,createTime) values ('0b01d176-d809-44fb-82fe-880b51605352',
'0b01d176-d809-44fb-82fe-880b51605352','1801','?????',
'1','2013-10-30 10:46:05');

……（省略完整代码）

　insert into Account_T(id,
ID,Accountcode,Accountname,
depth,createTime) values ('37c6c1f7-7e13-4729-96ac-24647e52354e',
'37c6c1f7-7e13-4729-96ac-24647e52354e','2251','?????',
'1','2013-10-30 10:46:05');

……（省略完整代码）

　insert into Account_T(id,
ID,Accountcode,Accountname,
depth,createTime) values ('3e00e4cf-2b44-4bf5-9735-3c6548c353bb',
'3e00e4cf-2b44-4bf5-9735-3c6548c353bb','1401','?????',
'1','2013-10-30 10:46:05');

insert into Account_T(id,
ID,Accountcode,Accountname,
depth,createTime) values ('78a2fcf1-5add-44e1-9644-e1370ffb7ec8',
'78a2fcf1-5add-44e1-9644-e1370ffb7ec8','1111','?????',
'1','2013-10-30 10:46:05');

　insert into Account_T(id,
ID,Accountcode,Accountnam
Fields:
```
      MT    Field     Offset           Type VT      Attr        Value Name
79332d70  4000096       4        System.Int32   1 instance      35374 m_arrayLength
79332d70  4000097       8        System.Int32   1 instance      35373 m_stringLength
79331804  4000098       c        System.Char    1 instance         32 m_firstChar
79330b24  4000099      10        System.String  0 shared        static Empty
    >> Domain:Value    000e34e0:069101d0 00105cc0:069101d0 10e6c008:069101d0 <<
79331754  400009a      14        System.Char[]  0 shared        static WhitespaceChars
    >> Domain:Value    000e34e0:06910728 00105cc0:02910a38 10e6c008:1f781680 <<
0:000> !objsize 03eb2f2c
sizeof(03eb2f2c) =      70764 (    0x1146c) bytes (System.String)
```

陆续查看了其他几个对象：

!do 03d16de0
!do 03d23288

　内容与上面完全相同，猜测可能有很多此字符串对象被申请，并重复存储。继续查了十几个对象，发现仍然有很多重复值。

使用!gcroot 命令查看这几个对象的引用关系：

DOMAIN(00105CC0):HANDLE(**Pinned**):25213fc:Root:0a91aaf8(System.Object[])->
0294d6c4(System.Collections.ArrayList)->
0bdcafb0(System.Object[])->
03eb2f2c (System.String)

DOMAIN(00105CC0):HANDLE(**Pinned**):25213fc:Root:0a91aaf8(System.Object[])->
0294d6c4(System.Collections.ArrayList)->
0bdcafb0(System.Object[])->
03d16de0 (System.String)

DOMAIN(00105CC0):HANDLE(**Pinned**):25213fc:Root:0a91aaf8(System.Object[])->
0294d6c4(System.Collections.ArrayList)->
0bdcafb0(System.Object[])->
03d23288 (System.String)

可以观察到，这些具有相同值的字符串来自同一个 ArrayList 对象。标注 Pinned 的字符串不会被 GC 垃圾线程回收掉。

看一下此 ArrayList 的结构：

0:000> !do 0294d6c4
Name: System.Collections.ArrayList
MethodTable: 79332b54
EEClass: 790ee62c
Size: 24(0x18) bytes
 (C:\WINDOWS\assembly\GAC_32\mscorlib\2.0.0.0__b77a5c561934e089\mscorlib.dll)
Fields:

MT	Field	Offset	Type	VT	Attr	Value	Name
793042f4	40008fb	4	System.Object[]	0	instance	0bdcafb0	_items
79332d70	40008fc	c	System.Int32	1	instance	50000	_size
79332d70	40008fd	10	System.Int32	1	instance	50028	_version
79330740	40008fe	8	System.Object	0	instance	029f0370	_syncRoot
793042f4	40008ff	1c4	System.Object[]	0	shared	static	emptyArray

\>> Domain:Value   000e34e0:0691298c 00105cc0:02913914 10e6c008:1f78266c <<

0:000> !objsize 0294d6c4
sizeof(0294d6c4) =   93356580 (0x5908224) bytes (System.Collections.ArrayList)

0:000> !do 0bdcafb0
Name: System.Object[]
MethodTable: 793042f4
EEClass: 790eda64
Size: 256016(0x3e810) bytes
Array: Rank 1, Number of elements 64000, Type CLASS
Element Type: System.Object
Fields:
None

ArrayList 有 64 000，截至此时已经使用了 50 000 个，即存储了 50000 个 object（里面又

包含string）。另外，一个ArrayList对象竟然占用了内存资源约93MB。

在2h前还Dump了一个文件，打开这个2h前Dump的文件（15:23时Dump），看一下它的内存情况。

看一下.NET堆大小：

```
0:000> !eeheap -gc
Number of GC Heaps: 2
------------------------------
Heap 0 (000e8990)
generation 0 starts at 0x28519afc
generation 1 starts at 0x282d41e8
generation 2 starts at 0x02910038
ephemeral segment allocation context: none
 segment    begin    allocated    size
02910000 02910038  0690f5e4 0x03fff5ac(67106220)
1a530000 1a530038  1e522bd8 0x03ff2ba0(67054496)
26e30000 26e30038  28739168 0x01909130(26251568)
Large object heap starts at 0x0a910038
 segment    begin    allocated    size
0a910000 0a910038  0c3f8368 0x01ae8330(28214064)
24e30000 24e30038  2636eb88 0x0153eb50(22276944)
Heap Size    0xc9220fc(210903292)
------------------------------
Heap 1 (000e99b0)
generation 0 starts at 0x2bd41da4
generation 1 starts at 0x2babb6c4
generation 2 starts at 0x06910038
ephemeral segment allocation context: none
 segment    begin    allocated    size
06910000 06910038  0a90e200 0x03ffe1c8(67101128)
1e610000 1e610038  2260fa30 0x03fff9f8(67107320)
2b010000 2b010038  2be98218 0x00e881e0(15237600)
Large object heap starts at 0x0c910038
 segment    begin    allocated    size
0c910000 0c910038  0d67d840 0x00d6d808(14080008)
22e30000 22e30038  23dd5e88 0x00fa5e50(16408144)
Heap Size    0xab993f8(179934200)
------------------------------
GC Heap Size    0x174bb4f4(390837492)
```

查看.NET堆中的对象：

```
0:000> !dumpheap -stat
No export dumpheap found
0:000> .load sos
0:000> !dumpheap -stat
------------------------------
```

```
Heap 0
total 1653442 objects
------------------------------
Heap 1
total 1437465 objects
------------------------------
total 3090907 objects
Statistics:
      MT        Count    TotalSize Class Name
   7ae3ac30       1           12 System.Drawing.ColorConverter
   7aa32410       1           12 System.Collections.Generic.SortedList`2+KeyList[[System.String, mscorlib],[System.String, mscorlib]]
   7a5ea814       1           12 System.ComponentModel.GuidConverter
   7a5e7038       1           12 System.Diagnostics.PerformanceCounterCategoryType
   7a5e6248       1           12 System.Diagnostics.OrdinalCaseInsensitiveComparer
   ……（省略完整代码）
   79332b54    199506     4788144 System.Collections.ArrayList
   79333274      8356     5979288 System.Collections.Hashtable+bucket[]
   793042f4    148550    13408584 System.Object[]
   000e82c0       394    50954996         Free
   79330b24   1466354   251176960 System.String
Total 3090907 objects
```

可见字符串对象仍然是占用内存最多的一个对象。

查看任意一个字符串对象的引用关系：

```
0:000> !gcroot 072f7b28
Note: Roots found on stacks may be false positives. Run "!help gcroot" for more info.
Scan Thread 10 OSTHread e1c
Scan Thread 14 OSTHread ab0
Scan Thread 15 OSTHread dbc
Scan Thread 16 OSTHread d10
Scan Thread 17 OSTHread bc0
Scan Thread 18 OSTHread cd8
Scan Thread 8 OSTHread e20
Scan Thread 3 OSTHread d6c
Scan Thread 5 OSTHread 9f8
Scan Thread 2 OSTHread e3c
Scan Thread 4 OSTHread 4a4
Scan Thread 22 OSTHread e80
Scan Thread 23 OSTHread 54c
Scan Thread 24 OSTHread 5fc
Scan Thread 25 OSTHread 670
Scan Thread 27 OSTHread fc0
DOMAIN(00105CC0):HANDLE(Pinned):25213fc:Root:0a91aaf8(System.Object[])->
0294d6c4(System.Collections.ArrayList)->
```

0bdcafb0(System.Object[])->
072f7b28(System.String)

发现仍然有同样大小的一个 ArrayList，如下：

```
0:000> !do 0294d6c4
Name: System.Collections.ArrayList
MethodTable: 79332b54
EEClass: 790ee62c
Size: 24(0x18) bytes
 (C:\WINDOWS\assembly\GAC_32\mscorlib\2.0.0.0__b77a5c561934e089\mscorlib.dll)
Fields:
      MT    Field   Offset                 Type VT     Attr    Value Name
  793042f4  40008fb      4         System.Object[]  0 instance 0bdcafb0 _items
  79332d70  40008fc      c            System.Int32  1 instance    50000 _size
  79332d70  40008fd     10            System.Int32  1 instance    50028 _version
  79330740  40008fe      8           System.Object  0 instance 029f0370 _syncRoot
  793042f4  40008ff     1c4        System.Object[]  0   shared   static emptyArray
     >> Domain:Value  000e34e0:0691298c 00105cc0:02913914 10e6c008:1faf9048 <<

0:000> !do 0bdcafb0
Name: System.Object[]
MethodTable: 793042f4
EEClass: 790eda64
Size: 256016(0x3e810) bytes
Array: Rank 1, Number of elements 64000, Type CLASS
Element Type: System.Object
Fields:
None
```

查看 ArrayList 中最后一个元素内容：

```
0:000> !do 1bd65988
Name: System.String
MethodTable: 79330b24
EEClass: 790ed65c
Size: 348(0x15c) bytes
 (C:\WINDOWS\assembly\GAC_32\mscorlib\2.0.0.0__b77a5c561934e089\mscorlib.dll)
String: 2013-10-30 13:18:07 [记载] TID:1 delete from [ARAD] where ID='369f826a-3f04-4686-a9f9-6887858cdc50' and Type='d6b6deeb-88fb-4e28-bacc-a8f2bb3b449c'
Fields:
      MT    Field   Offset                 Type VT     Attr    Value Name
  79332d70  4000096      4            System.Int32  1 instance      166 m_arrayLength
  79332d70  4000097      8            System.Int32  1 instance      165 m_stringLength
  79331804  4000098      c             System.Char  1 instance       32 m_firstChar
  79330b24  4000099     10          System.String  0   shared   static Empty
     >> Domain:Value  000e34e0:069101d0 00105cc0:069101d0 10e6c008:069101d0 <<
  79331754  400009a     14         System.Char[]  0   shared   static WhitespaceChars
     >> Domain:Value  000e34e0:06910728 00105cc0:02910a38 10e6c008:1faf805c <<
```

最开始分析的 Dump 文件（17:05 时 Dump）中，ArrayList 中最后一个元素内容：

0:000> !do 1bcca510
Name: System.String
MethodTable: 79330b24
EEClass: 790ed65c
Size: 348(0x15c) bytes
 (C:\WINDOWS\assembly\GAC_32\mscorlib\2.0.0.0__b77a5c561934e089\mscorlib.dll)
String: 2013-10-30 13:18:07 [记载] TID:1 delete from [ARAD] where ID='369f826a-3f04-4686-a9f9-6887858cdc50' and Type='d6b6deeb-88fb-4e28-bacc-a8f2bb3b449c'
Fields:
MT	Field	Offset	Type	VT	Attr	Value	Name
79332d70	4000096	4	System.Int32	1	instance	166	m_arrayLength
79332d70	4000097	8	System.Int32	1	instance	165	m_stringLength
79331804	4000098	c	System.Char	1	instance	32	m_firstChar
79330b24	4000099	10	System.String	0	shared	static	Empty
>> Domain:Value	000e34e0:069101d0	00105cc0:069101d0	10e6c008:069101d0 <<				
79331754	400009a	14	System.Char[]	0	shared	static	WhitespaceChars
>> Domain:Value	000e34e0:06910728	00105cc0:02910a38	10e6c008:1f781680 <<				

两个 Dump 文件中的 ArrayList 中最后一个元素完全相同，都是记录 13:18:07 时的一个 SQL。随后比较 ArrayList 中的另外两个元素，发现存储的值也都相同，另外，还可以推断 w3wp 在 15:23~17:05 之间没有发生回收或异常。

经确认，此 ArrayList 对象用于缓存业务操作日志，每隔 7s 会把内存中的内容写入文件然后清空。其代码如下：

```
//伪代码
private void WriteLog(object obj)
{
    try
    {
        while (true)
        {
            lock (writeLock)
            {
                writeLogs(logFileName);

                writeStats(statFileName);

                System.Threading.Thread.Sleep(new System.TimeSpan((System.Int64)10000 * 7000));
            }
        }
    }
    catch { }
}
```

```
private void LogFile(string fileName)
    {
        try
        {
            //把内存对象写到文件中
        }
        finally
        {
            list.Clear();   //清空 ArrayList
        }
        catch {
            ……（省略完整代码）
        }
    }
```

正常情况下，每隔 7s ArrayList 对象内容会被存储到文件中，之后再被清空一次。

而两个 Dump 文件之间间隔近 2 个小时，内容完全相同，说明 ArrayList 有两个小时没有清理了。推测上面代码很可能发生了异常。

下面是异常检测分析过程，看一下 w3wp.exe 进程中，所有进程状态：

0:000> !threads
ThreadCount: 15
UnstartedThread: 0
BackgroundThread: 15
PendingThread: 0
DeadThread: 0
Hosted Runtime: no

	ID	OSID	ThreadOBJ	State	PreEmptive GC	GC Alloc Context	Lock Domain	Count	APT	Exception
	10	1	e1c	000e4e40	1808220	Enabled	00000000:00000000	000e34e0	0	MTA	(Threadpool Worker)
	14	2	ab0	000eb0d8	b220	Enabled	00000000:00000000	000e34e0	0	MTA	(Finalizer)
	15	3	dbc	00102070	80a220	Enabled	00000000:00000000	000e34e0	0	MTA	(Threadpool Completion Port)
	16	4	d10	00105580	1220	Enabled	00000000:00000000	000e34e0	0	Ukn	
	17	5	bc0	0014e5d0	380b220	Enabled	3643593c:36436c80	00105cc0	1	MTA	(Threadpool Worker) **System.UnauthorizedAccessException** (36435510)
	18	6	cd8	0012a1a8	180b220	Enabled	2b2a4b84:2b2a5834	000e34e0	0	MTA	(Threadpool Worker)
	8	7	e20	0f639038	880a220	Enabled	00000000:00000000	000e34e0	0	MTA	(Threadpool Completion Port)
	3	8	d6c	0f617bf8	220	Enabled	00000000:00000000	000e34e0	0	Ukn	
	5	9	9f8	129df678	220	Enabled	00000000:00000000	000e34e0	0	Ukn	
	2	a	e3c	10ecb640	220	Enabled	00000000:00000000	000e34e0	0	Ukn	

4	b	4a4	10de0ad8	220	Enabled	00000000:00000000	000e34e0	0	Ukn
22	d	e80	10f00008	180b220	Enabled	00000000:00000000	000e34e0	0	MTA (Threadpool Worker)
23	c	5fc	10eb6e30	180b220	Enabled	361ffaa8:362007b0	000e34e0	0	MTA (Threadpool Worker)
24	10	670	00186b08	180b220	Enabled	00000000:00000000	000e34e0	0	MTA (Threadpool Worker)
27	11	1e8	129d78c0	220	Enabled	00000000:00000000	000e34e0	0	Ukn

可以看到有线程产生了 System.UnauthorizedAccessException 异常，并且这个线程 ID 是 17。切换到 17 号线程，看一下当前堆栈发生的异常信息：

```
0:000> ~17s
eax=00000000 ebx=00000000 ecx=00000000 edx=00000000 esi=00000000 edi=0f5df5e8
eip=7c9585fc esp=0f5df5a8 ebp=0f5df610 iopl=0         nv up ei pl nz na pe nc
cs=001b  ss=0023  ds=0023  es=0023  fs=003b  gs=0000             efl=00000206
ntdll!KiFastSystemCallRet:
7c9585fc c3              ret
0:017> !pe
Exception object: 36435510
Exception type: System.UnauthorizedAccessException
Message: 对路径 "C:\Program Files\King\**System\logs\Operator.log" 的访问被拒绝。
InnerException: <none>
StackTrace (generated):
    SP       IP       Function
    0F5DF618 799D0040 mscorlib_ni!System.IO.__Error.WinIOError(Int32, System.String)+0x75bf20
    0F5DF674 792EB35B mscorlib_ni!System.IO.FileStream.Init(System.String, System.IO.FileMode, System.IO.FileAccess, Int32, Boolean, System.IO.FileShare, Int32, System.IO.FileOptions, SECURITY_ATTRIBUTES, System.String, Boolean)+0x48b
    0F5DF76C 792EA9B2 mscorlib_ni!System.IO.FileStream..ctor(System.String, System.IO.FileMode, System.IO.FileAccess, System.IO.FileShare, Int32, System.IO.FileOptions)+0x42
    0F5DF794 7927786F mscorlib_ni!System.IO.StreamWriter.CreateFile(System.String, Boolean)+0x3f
    0F5DF7A8 7927780B mscorlib_ni!System.IO.StreamWriter..ctor(System.String, Boolean, System.Text.Encoding, Int32)+0x3b
    0F5DF7C8 792777BE mscorlib_ni!System.IO.StreamWriter..ctor(System.String, Boolean, System.Text.Encoding)+0x1e
    0F5DF7DC 13FF7E38 King_T_eap_AopBase!King.KAP.Logging.LogManager.writeToFile(System. Collections.IList, System.String)+0x48

StackTraceString: <none>
HResult: 80070005
```

可以看到，在调用 LogManger.writeToFile 方法中的 CreateFile 方法时，产生了文件拒绝访问的异常。

继续，看一下当前堆栈中调用的对象：

```
0:017> !dso
```

```
OS Thread Id: 0xbc0 (17)
ESP/REG    Object    Name
0f5df5fc 36435510 System.UnauthorizedAccessException
0f5df630 364352e0 System.Security.Permissions.FileIOPermission
0f5df750 36435064 System.IO.StreamWriter
0f5df79c 06961618 System.String      C:\Program Files\King\**System\logs\Operator.log
0f5df7c0 0693adb0 System.Text.DBCSCodePageEncoding
0f5df7cc 36435064 System.IO.StreamWriter
0f5df7e4 02dad738 System.Collections.ArrayList
0f5df818 06961618 System.String      C:\Program Files\King\**System\logs\Operator.log
0f5df820 02dad738 System.Collections.ArrayList
0f5df858 0294f7b4 System.Object
0f5df85c 0294f7d8 King.KAP.Logging.LogManager
0f5df8a4 0294f804 System.Threading._ThreadPoolWaitCallback
```

经确认，该日志文件大小达到 500M 后会自动备份，并新建一个新文件来存储。这些操作全部是后台调度服务来完成的，且是系统日志模块自身出现故障，所以系统并不能记载错误异常信息。如果不用这样的进程调试工具，一般的调试工具很难检测到。

解决方案：
为应用程序设置对 logs 文件夹的写权限。

第6章
数据库性能分析及故障诊断方法

本章内容

- 数据库优化概述
- 效率专题研究
- 优化方法指令
- 性能故障检测方法

6.1 数据库优化概述

一般开发工程师只要了解 SQL 语法基本上就可以进行开发任务了，但优化工程师除了要掌握这些，还要了解数据库的各个层面，如：存储引擎、编译优化、大并发下的阻塞/死锁诊断，及 CPU/内存/磁盘负载下的各种诊断方法。除些之外，优秀的设计也是数据库高性能的前提保障。

如我们接到客户反馈说：系统使用起来非常慢，并且技术人员已经初步定位到是数据库的原因，即使这样，像 CPU、编译、内存、I/O、tempdb、阻塞/死锁、存储等任何一个部件出现瓶颈都会导致客户描述的问题，那接下来应该怎么做呢？

本章的内容就是作者在解决客户问题过程中，积累的一些常用诊断经验。

首先分享四个效率专题研究案例，这些案例主要从设计角度来介绍一些构建高性能数据库的经验。接下来，会对性能优化方法及常用工具做个介绍。最后，会对 SQL Server 导致的一些常见的 CPU、内存、I/O、tempdb、阻塞/死锁等故障的监控方法进行介绍。

6.2 效率专题研究

6.2.1 数据库无法收缩变小原因分析案例汇总

数据库在使用过程中会逐渐增大，我们遇到比较棘手的问题是收缩数据库或文件，但空间仍无法释放，结果数据库变得越来越庞大，出现严重的效率问题，也不好管理。

汇总一下目前我们遇到的问题，大致分为以下两类：

（1）未使用空间无法释放（unused）——客户普遍存在的问题。

首先是江西九江客户遇到的大数据库无法收缩问题，当时客户使用三个月达到 27GB 数据（实际数据只有 600MB，其余为数据/索引碎片）。这个问题比较普遍，很多客户都有这种问题。

（2）未分配空间无法释放（unallocated space）——个别客户遇到的问题。

郑州某客户遇到过大数据库无法收缩问题，数据库 47GB 中，仅有 8GB 为数据真实空间大小，其余均为未分配空间。

如图 6-1 所示为数据文件中的空间分布图，带批注的矩形为产生问题的两个点。

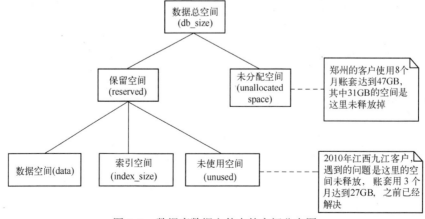

图 6-1　数据库数据文件中的空间分布图

这两个问题原因完全不同，相对于 unused 空间释放来说，unallocated space 空间释放其实更容易。因为未分配空间（unallocated space）是分配给数据库但还未分配到具体的对象（如表或索引等）使用；而未使用空间（unused）是已经被分配到具体对象了，涉及数据/索引碎片问题，要进行特殊处理后才能回收。本案例中，该问题是由客户数据库的数据文件页损坏所致。而 unused 空间释放更复杂一些，需要对表的设计进行修改才能解决。

1．unused space 无法释放原因定位

接到江西九江客户 SOS 邮件，说系统使用很慢，经过现场定位发现是数据库的问题，并且数据库非常大且不能使用 SSMS 正常收缩。另外一个现象是磁盘灯一直闪烁，队列很大。

当时现场支持后，我只带了一个 17GB 的客户数据库回北京总部分析，此时客户数据库已经增大到 27GB 了。对这个数据库做了一下研究分析，过程如下：

数据库空间分配在哪？如图 6-2 所示。

	reserved	data	index_size	unused
1	17660824 KB	2216280 KB	97448 KB	15347096 KB

图 6-2　数据库空间分配

在保留空间 17GB 中，数据占用了 2GB，索引占用空间不到 100MB，unused 占了 15GB。也就是说数据库 17GB 中有 15GB 是用作 unused 的。在数据库空间分布中，只能得到 unused 占用了绝大多数空间。

又看了一下各个表的空间分布，每个表中的 unused 占用都比较大，而且是操作频率越高的表 unused 值越大。unused 其实就是数据碎片和索引碎片。

之后尝试使用 SSMS 的数据库或文件收缩功能，只收缩了 1GB，数据库还是比较大。

又看了下数据库所有表的碎片，如图 6-3 所示。

TSOFT\...\SQL优化\1.sql　TSOFT\...\SQL优化\2.sql　TSOFT\...\SQL优化\3.sql*　对象资源管理器详细信息

```
select b.avg_fragmentation_in_percent as 索引碎片X,g.avg_fragmentation_in_percent as 索引碎片Y,b.fragment_cour
--b.object_id, b.index_id, b.index_type_desc, alloc_unit_type_desc,index_deph
from IndexFragment_Bad b inner join IndexFragment_Good g
on b.object_id = g.object_id and b.index_id = g.index_id
where (b.avg_fragmentation_in_percent <> 0 OR g.avg_fragmentation_in_percent <> 0)
```

结果　消息

	索引碎片X	索引碎片Y	碎片数量X	碎片数量Y	平均碎片中页面数X	平均碎片中页面数Y
1	99.9043977055449	11.8181818181818	4184	15	1.0019120458891	57.7333333333333
2	99.833748960931	0	1203	1	1.00083125519534	253
3	99.8322147651007	6.25	596	2	1	57
4	99.825	0	4000	1	1.00025	171
5	99.7448979591837	7.69230769230769	394	2	1.01776649746193	46.5
6	99.7174143412222	9.52380952380952	2831	3	1.00035323207347	50
7	99.604743083004	0.99009900990099	253	10	1	20.2
8	99.5938628158845	10.2040816326531	4432	6	1.00022563176895	65
9	99.4736842105263	1.46341463414634	190	17	1	12.0588235294118
10	99.4252873563218	3.74331550802139	174	26	1	7.19230769230769
11	99.4134897360704	1.87265917602996	341	15	1	17.8
12	99.3710691823899	1.62162162162162	159	17	1	10.8823529411765
13	99.3031358885017	12.1771217712177	287	69	1	3.92753623188406
14	99.3017009847807	10	5585	21	1.00662488809311	71.3809523809524
15	99.2857142857143	15.4676258992806	280	82	1	3.39024390243902
16	99.2857142857143	2.11864406779661	280	14	1	16.8571428571429
17	99.250936329588	25	267	4	1	20.5
18	99.2424242424242	31.6742081447964	132	130	1	1.7
19	99.2094861660079	0	253	8	1	29.125
20	99.2075471698113	1.86092066601371	2648	93	1.00075528700906	21.9569892473118
21	99.1935483870968	0	124	1	1	85
22	99.1735537190083	1.68776371308017	242	16	1	14.8125
23	99.163179916318	2.92887029288703	239	20	1	11.95
24	99.1596638655462	37.0689655172414	238	222	1	1.56756756756757
25	99.1189427312775	0.327868852459...	453	3	1.00220750551876	101.666666666667
26	99.0990990990991	0.813008130081...	111	8	1	15.375
27	99.0867579908676	0	219	6	1	24.8333333333333
28	99.0740740740741	3.40909090909091	108	8	1	11
29	99.0277203144394	3.67180417044424	4834	313	1	14.0958466453674
30	99.0277203144394	0	4834	NULL	1	NULL
31	98.9473684210526	9.69387755102041	190	44	1	4.45454545454545
32	98.936170212766	2.9045643153527	282	24	1	10.0416666666667

图 6-3　查看数据库表碎片

据统计，索引碎片超过50%的索引个数近200个，其中索引碎片超过90%的有113个。一般索引碎片可能在10%以下为最佳，当超出10%时，建议整理索引（碎片<30%）或重建索引（碎片>30%）。还有，以上索引中，聚集索引碎片严重的占90%中的大多数。

基于此数据结论，我对数据库做了碎片整理。整理后再收缩数据库，发现数据库文件还是没有收缩下来。经过进一步研究，发现了原因。

问题原因：

表按存储方式分为两种类型：

◆ 聚集表：创建聚集索引的表。

◆ 非聚集表（堆）：未创建聚集索引的表。

对于聚集表，通过重新创建/组织可以起到对表数据碎片清理的目的，因为聚集索引在等二叉树的叶节点即是表数据位置。而对于堆，由于没有聚集索引，数据存储相对比较散列，即使碎片整理也只能对所有非聚集索引进行整理，不能对数据进行整理。

如果表不含有聚集索引，即使采用 SSMS 的数据库或文件收缩功能，也收缩不掉，导致数据库太大，不好管理，查询速度变慢。

补充：聚集索引创建的必要性。

◆ 我做了一个实验，一旦创建聚集索引后再删除聚集索引，查询计划会由聚集搜索变为原来的表扫描，但扫描成本仍然为 15%，而不是之前的 41%。换句话说，由于聚集索引能够保证数据物理存储是连续的，则查找大数据量时一般会比较快，磁盘轴每次物理寻道（移动）或逻辑 lookup/rid 操作时间较短；如果表中没有聚集索引，则可能数据是分散存储的，用一般的表扫描可以使磁盘轴每次寻道（移动）时间或逻辑 lookup/rid 操作比较长，整个表扫描时间就会长很多。

◆ 聚集索引也可以避免表中产生数据碎片，建议表中都保留一个聚集索引。产生数据碎片的主要原因是表中缺少聚集索引，如果表中含有聚集索引，可以使数据在任何更新操作（插入/删除）后自动移动数据保证数据物理连续，很少会产生数据存储间的空闲碎片；否则，不久就会产生很多碎片，因为聚集索引的索引结构的叶索引就是数据位置，这些是非聚集索引做不到的。解决办法是，在表中保留一个聚集索引，在更新（插入/修改/删除）时就避免数据碎片，而不是等到账套大了再缩小。那些不含有聚集索引且插入更新频繁的表是最容易导致体积变大的。建议使用 CreateTime，用户插入单据时一般是按时间的，按索引排序原则会插入到数据最后，这样也不会导致数据匹配移动。尽量不要用 ID（Guid），由于 Guid 值大小是随机的，这样的字段建立聚集索引经常会导致一个很小的插入也会引起大批量数据移动。用 CreateTime 做聚集索引也会为以后数据量很大时，做数据分区提供方便。

◆ 最重要的一点：虽然聚集索引扫描与表扫描查询速度在很多时候是一样的，但聚集索引查找是所有操作中最快的，比非聚集索引查找还快。

客户问题解决方案：

（1）结合业务设计尽量确保每个表都含有一个聚集索引。

（2）每周做调度服务：索引碎片整理（碎片<30%）或 索引重建（碎片>30%）。

最后，客户数据库文件由 27GB 变为了不到 1GB 大小，而 1GB 数据库文件中的数据

(DataSize)部分仅为590MB。另外,性能也大幅提升,磁盘也已经不存在队列。

2. Unallocated Space 无法释放原因定位

客户问题:

客户使用了8个月,数据库达到47GB;使用SSMS无法进行完全收缩,收缩后数据库文件由47GB变为42GB,仅收缩掉5GB空间,仍然比较大,给客户维护(如备份)带来不便。

问题原因:

对这个数据库进行了研究。这个47GB的数据库中数据文件空间分布(sp_spaceused command)如图6-4所示。

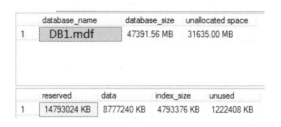

图6-4 数据库文件空间分配

Database_size:数据库总空间;unallocated space:未分配空间;reserved:已分配空间;Data:已分配空间中的对象占用的空间;index_size:已分配空间中的索引对象占用的空间;unused:已分配空间中还未被使用的空间。

数据库47GB总空间中,数据占用空间(reserved)仅有14GB,大部分空间为未分配空间(Unallocated Space),占用了31GB。**而这里的Unallocated Space空间占用了总数据库空间的70%左右,应当释放这个空间。**

尝试使用收缩命令对数据文件进行收缩:

```
DBCC SHRINKFILE (MsSql , 15000)     --MsSql 为数据文件的逻辑文件名
GO
DBCC SHRINKFILE (MsSql , 0, TRUNCATEONLY)
GO
```

执行过程中有如下错误产生:

DBCC SHRINKFILE: 无法移动页 **1:5362183**,因为该页所属的分区已删除。

(1 行受影响)
DBCC 执行完毕。如果 DBCC 输出了错误信息,请与系统管理员联系。

(1 行受影响)
DBCC 执行完毕。如果 DBCC 输出了错误信息,请与系统管理员联系。

shrink 操作在回收 31GB 未分配空间(Unallocated Space)时,回收到近 5GB 时就发生了错误,导致不能继续收缩剩下的 26GB 空间。上面报错误应该是客户的数据库文件中的区已经出现了错误(shrink 操作是以区为单位进行操作的,一个区相当于 8 页)。

尝试把数据库的这个数据文件放到同文件组下的其他文件。

在数据库的"属性"窗口→"文件"选项中增加一个数据文件 DB2.mdf。

在数据库的"任务"→"收缩"→"文件"中,选择"通过将数据迁移到同一文件组中的其他文件来清空文件"选项把损坏的数据文件 DB1.mdf 的内容放到新的数据文件 DB2.mdf 中。

但执行过程中,又报错误了,如图 6-5 所示。

图 6-5 报某一数据页无法移动

图 6-5 中 MsSql(逻辑文件名)文件的 1362183 页无法移动,在前面收缩文件时也是报告这一数据页有问题。

错误页分析:

收缩数据文件和在文件组内移动数据都报同一页 pageid=5362183 出错,如图 6-6 所示。

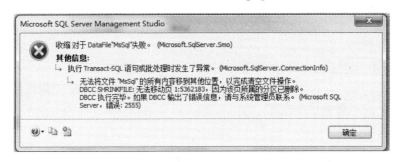

图 6-6 报某一数据页无法移动

DBCC SHRINKFILE: 无法移动页 1:5362183,因为该页所属的分区已删除。

(1 行受影响)
DBCC 执行完毕。如果 DBCC 输出了错误信息,请与系统管理员联系。

(1 行受影响)
DBCC 执行完毕。如果 DBCC 输出了错误信息,请与系统管理员联系。

这个错误尝试使用 SQL Server 的 DBCC CHECKDB 检查不到:

CHECKDB 在数据库'DB1' 中发现 0 个分配错误和 0 个一致性错误。
DBCC 执行完毕。如果 DBCC 输出了错误信息,请与系统管理员联系。

不能用数据库的修复机制进行修复。用下面的 DBCC PAGE 看一下这页的内容,看能否得到这页是哪个表或索引的页。

执行命令：

DBCC TRACEON(3604,-1)
DBCC PAGE(8,1,5362183,3)

运行结果如下：

DBCC 执行完毕。如果 DBCC 输出了错误信息，请与系统管理员联系。
PAGE: (1:5362183)
BUFFER:
BUF @0x0000000088FAE940
bpage = 0x000000008826E000   bhash = 0x0000000000000000   bpageno = (1:5362183)
bdbid = 8   breferences = 0   bcputicks = 0
bsampleCount = 0   bUse1 = 43374   bstat = 0xc00009
blog = 0x32159   bnext = 0x0000000000000000
PAGE HEADER:
Page @0x000000008826E000
m_pageId = (1:5362183)   m_headerVersion = 1   m_type = 1
m_typeFlagBits = 0x0   m_level = 0   m_flagBits = 0x220
m_objId (AllocUnitId.idObj) = 1443501   m_indexId (AllocUnitId.idInd) = 256
Metadata: AllocUnitId = 72057688639209472   **Metadata: PartitionId = 0**
Metadata: IndexId = -1   **Metadata: ObjectId = 0**   m_prevPage = (1:5362182)
m_nextPage = (1:5362184)   pminlen = 869   m_slotCnt = 8
m_freeCnt = 696   m_freeData = 7480   m_reservedCnt = 0
m_lsn = (115447:11760:474)   m_xactReserved = 0   m_xdesId = (0:15618059)
m_ghostRecCnt = 0   m_tornBits = -575686618
Allocation Status
GAM (1:5112320) = ALLOCATED   SGAM (1:5112321) = ALLOCATED
PFS (1:5354256) = 0x40 ALLOCATED 0_PCT_FULL   DIFF (1:5112326) = NOT CHANGED
ML (1:5112327) = NOT MIN_LOGGED

比较遗憾，无法得到这页相关的信息：

Metadata: ObjectId = 0   //无法得到这页是属于哪个表，不管是表对象还是索引对象都应该返回一个有效的值。

Metadata: PartitionId = 0   //无法得到这页是属于哪个分区，我们没有做分区，这里应该所有表都显示默认分区的 PartitionId 才对，但这里丢失了。

```
Metadata: IndexId = -1        //这里 sql 无法判断出是数据还是索引,干脆返回了-1;
                              //页不损坏的情况下是要返回以下三个值中的其中一个:
                              // --IndexId=0  堆  数据页。
                              //--IndexId=1  聚集索引   索引页。
                              //--IndexId>1  非聚集索引   索引页。
```

这一页不属于任何表,不属于任何分区,在文件中是一个死页。

如果是表 T_A 中的页或某个索引 I_id 出了问题,则只要创建一个新表并把这些数据导入到新表,然后删除旧表,重命名新表为旧表名称就可以解决坏页问题了。并且不用在数据库之间导全部数据,只要在同一个数据库下的不同文件组就可以(新建立一个文件组,把坏页所在的对象如表数据导入新文件组)。

但这里什么信息也没有,无法找到损坏页的归属对象,只好对所有数据导入到新库,利用新的数据文件、数据库导入数据。

这一次成功了,前面方法失败的原因应该是访问数据的方式不是复制。

步骤如下:

(1)备份 DB1 整个数据库。

(2)把 DB1 数据库重命名为 DB1_orinal。

(3)创建一个新的数据库 DB1,使用 SSMS 把 DB1_orinal 库的结构和数据全部导入到 DB1 数据库中。

数据导完后,再次使用前面步骤中的 shrink 操作,已经不再报之前的数据页错误了。以后就可以使用 SSMS 的"文件收缩"功能进行正常的收缩操作了。

收缩后,使用的数据空间仅为 8GB(之前是 47GB),而物理文件也正好是 8G;Unallocated Space 也只有 32.97MB 了(之前是 31GB)。

6.2.2　数据库碎片增长过快原因分析及建议方案

碎片是数据库始终困扰 DBA 的一个问题,因为它不仅导致查询变慢,也会导致数据库体积变大不好管理。数据库碎片与磁盘碎片原理相似,不可能完全避免,但遵循一定的设计原则,可以控制碎片增长的速度。

如下是几点我在工作中实际遇到的问题和解决方案。

1．每个表要保留一个聚集索引

表按存储方式可分为以下两种类型:

● 聚集表:创建聚集索引的表。

● 堆表(非聚集表):未创建聚集索引的表。

对于聚集表,可以通过重新创建或重新组织起到对表数据碎片清理的作用,因为聚集索引在等二叉树的叶节点即是表数据位置。而对于堆表,由于没有聚集索引,数据存储相对比较散列,即使碎片整理也只能对所有非聚集索引进行整理,不能对数据进行整理。

如果表不含有聚集索引,即使采用 SSMS 的数据库或文件收缩功能也收缩不掉,导致数据库太大,不好管理,查询速度变慢。

2．不要用可能会被再次修改的字段做聚集索引

我们建议用"时间"字段作为聚焦索引。聚集索引字段用 CreatedTime 这种永远不改变的字段来代替 UpdatedTime 这种后续可能被修改的字段。

由于索引的规则数据库要对字段重新排序，一旦聚集索引字段被修改，就会导致数据的批量移动，更新效率较低，也容易导致碎片产生。

3．Guid 类型字段建立索引后会产生碎片较多

- ◆ 对表的主外键关系（主键和外键）都不要用 Guid 类型标识，一旦使用必然会创建索引。
- ◆ 对 Guid 类型的字段尽量不建议创建索引。
- ◆ Guid 类型字段应该禁止建立聚集索引。

由于 Guid 值大小是随机的，使用这样的字段建立聚集索引时，即使一个很小的插入也会导致大量的批量数据移动，所以有人形容 Guid 就像一只爱跳的猴子。

Guid 类型字段除了大小的不确定性以外，它本身占用的字节也比较大，会导致索引在很短时间内就产生大量碎片。

如果产品中违反了以上规则，却已经在使用了，可以通过如下方式进行优化：

- ◆ 数据库端修正

SQL Server 有内置的函数：NEWSEQUENTIALID()；目前默认的函数为 NEWGUID()，是无序的，从而避免在插入数据时索引碎片的产生。

- ◆ 代码端生成修正

根据 Guid+服务器网卡 ID+时间，自己代码生成有序的 Guid。

通过以下几个例子分析 Guid 类型字段对碎片的影响：

- ◆ 对比 int 和 guid 对碎片的影响

插入 10 万条数据后，碎片情况如图 6-7 所示。

图 6-7　碎片情况

Guid 类型的索引碎片约为 99%，而 Int 类型的索引碎片仅为 2%左右。

- ◆ 对比有序生成 Guid 和无序生成 Guid 对碎片的影响

同样，插入 10 万条数据，碎片情况如图 6-8 所示。

图 6-8　碎片情况

进行随机插入操作后，T1 和 T2 表碎片有了明显的差异，有序 Guid 索引产生的碎片大概是无序 Guid 索引碎片的一半。

在数据空间占用方面，采用有序 Guid 方式仅占用了 71 页，而无序 Guid 方式占用了 101 页。

4．尽量用系统的基本类型表示相应的业务字段类型

考虑到整体的一致性，有时一些业务字段不用系统基本类型表示。举一个实际的例子，某产品中没有用 DateTime 基本日期类型字段表示日期，而是用了 char(20)来表示，实验数据如图 6-9 所示。

表名	表数据量	索引名	索引类型	页填充度	索引碎片	页数
T2	99999	I_CreateTime2	CLUSTERED INDEX	61.795997034...	69.292742050...	14026
T1	99999	I_CreateTime2	CLUSTERED INDEX	94.561984185...	0.6159938400...	9091

图 6-9　碎片情况

顺序插入 10 万条数据后,聚集索引列类型为 char(19)的索引碎片增加到 69%;而 DateTime 列的索引并没有增加,说明 char(19)类型列作为索引会导致大量碎片的产生。

因此要尽量用系统的基本类型表示相应的业务字段类型。

5. 数据库碎片归纳和总结

除此之外,我还做了其他数据类型对碎片影响的对比实验。这些实验结论仅支持该实验环境下的数据。首先,介绍一下实验中的几个概念。

顺序插入:是指往表中追加数据,表可能是空的,也可能是有数据的。

随机插入:是指按一定算法往现有数据分布中穿插数据,打乱现有数据顺序。

随机更新:是指按一定算法随机修改表中的数据。

删除:按一定算法随机删除现有比例的数据。

(1)实验结论

1)【聚集表】VS【堆表】

顺序插入数据时,非聚集索引产生的碎片很少。

不管是聚集表还是堆表中的非聚集索引,插入对碎片的影响基本是一致的(30%)。

不管在聚集表还是堆表中,更新对碎片的影响非常大(99%)。

删除不会产生额外的碎片。

2)【char(19)】 VS 【DateTime】

顺序插入数据时,char(19)列聚集索引会增加大量碎片(70%),而 DateTime 列聚集索引几乎不产生碎片。

随机插入数据时,char(19)和 DateTime 都会产生碎片(27%)。

更新时,char(19)和 DateTime 都会产生较多碎片,char(19)达到 80%,DateTime 达到 90%。

删除不会产生额外的碎片。

3)【char(19)】VS【Int】

顺序插入数据时,聚集列类型为 int 的索引没有产生碎片;但聚集列类型为 char(19)的索引产生了 25%的碎片。

随机插入数据时,char(19)的列产生了 99%的碎片;而 Int 类型的列仅产生 18%的碎片。

删除不会产生额外的碎片。

4)【聚集索引】VS【非聚集索引】

顺序插入数据时,对这两个索引都是相同的,聚集索引和非聚集索引的碎片差别不是很大;最主要的还是索引列数据类型,如 char(19)会产生很多碎片(>50%),而 DateTime 几乎不产生碎片(<2%)。

随机插入数据时,数据类型 char(19)比 DateTime 类型更容易产生碎片;另外,在表格中同样是 char(19)类型,聚集索引(99%)比非聚集索引更容易产生碎片(67%)。

更新数据:更新聚集索引,会对所有非聚集索引进行更新;更新非聚集索引则仅对这一个索引进行修改。所以上表格中 T2(I_Code)索引并没有产生碎片,因为它没有被更新。从其他三个索引碎片程度可以看到,更新对碎片影响也是较大的(>90%)。

删除不会产生额外的碎片。

5)【char(50)】VS【char(10)】

顺序插入数据时,char(50)列索引碎片(55%)明显比 char(10)碎片(25%)要大。索引列字段占用字节越大,越容易产生碎片。

随机插入、更新等操作碎片情况基本一致。

6)【Int】VS【Guid】

顺序插入数据时,Guid 类型的索引碎片为 99%,而 Int 类型的索引碎片仅 2%多一点。

7)【char(100)】VS【varchar(100)】

在顺序插入数据(新增数据)时,相同字节的 char 比 varchar 更容易生成碎片,char 比 varchar 大概多生成 50%的碎片。

在随机插入数据和随机更新时,也是如此。相同字节的 char 比 varchar 大概多生成 50%的碎片。

(2)归纳小结

顺序插入:一般数据类型,如 Int, DateTime 等不会产生碎片;char(19)类型会产生较多的碎片;Guid 类型更容易产生碎片,除了长度以外,更主要原因是它产生值的不确定性;索引列字段占字节越大,产生的索引碎片越大。

随机插入:所有数据类型都易产生碎片;Char(19)这样的数据字段索引比基本类型(Int、DateTime 等)更易产生碎片;聚集索引比非聚集索引更易产生碎片。

随机更新:更新聚集索引列消耗的时间更长;随机更新比随机插入产生更多的碎片,在聚集表或堆表都是如此;更新聚集索引比更新非聚集索引代价更大,因为更新聚集索引会对所有非聚集索引进行更新,而更新非聚集索引仅对这一个索引进行修改。

删除操作:对碎片影响非常小,可以忽略。

6.2.3 聚集索引对插入效率的影响

经常听到有经验的数据库开发人员说,表增加了聚集索引后会影响插入效率,真是这样吗?结论是:可能会变慢,也可能会变快,这要看具体应用场景。

我做了一个实验,分别对 Guid 和 Int 字段创建聚集索引及不创建聚集索引 3 种情况进行了个测试。

表默认数据量 10 万条,然后分别插入不同数据量数据,进行比较,测试结果如表 6-1 所示。

表 6-1 执行时间统计

记录数	执行时间/s		
	聚集索引(ID Int, not null, 自增)	无聚集索引(not null)	聚集索引(ID Guid, not null)
100	小于 1	小于 1	小于 1
1000	1	5	11
5000	12	27	59
10000	15	41	83

测试结果显示聚集索引(ID Int, not null, 自增)是最快的,而聚集索引(ID Guid, not null)

则是最慢的,并且分别插入 1000/5000/100000 条记录时,插入时间也是按递增比例规律增大。

原因分析:

并不是加了索引,插入或更新速度一定会慢。使用聚集索引（ID Int, not null, 自增）列作为聚集索引插入速度最快,是因为把这列设置为了递增,在插入时它不需要比较现有数据（因为数据库认为递增的列永远没有重复）,判断是否插入重复,直接插入到最后即可。不加任何聚集的场景不是最快的,是由于列设置成了唯一且非空的,在每条记录插入时 SQL Server 要对其进行唯一性校验、非空校验等逻辑操作。而聚集索引（ID Guid, not null）最慢是因为,Guid 每次生成值的大小都是随机的,而索引要求插入数据后维持排序原则,所以插入的新行是随机插入到现在数据位置中的,也就会导致数据值移动,这是最严重的情况,要尽量避免。

聚集索引应该选择在什么字段上创建呢?首先,我的建议是自增的 Int 是最快的。其次,聚集索引尽量建立在日期或时间戳列,也是很好的选择。

6.2.4 多表连接方案效率评估

1. 自然连接与左连接效率对比

SQL Server 逻辑连接分为全连接（full join）、左/右半连接（left join/right join）以及自然连接（inner join）等。

SQL Server 物理连接分为循环嵌套连接、哈希连接和合并连接。

从效率方面考虑,一般能用半连接实现的就不要用全连接;同样,能够用自然连接实现的就不要用半连接。但是,极少数特殊情况下执行结果可能与本规则相反。

下面是微软 MSDN 上的一段话:

> 避免 left join 和 null
>
> left join 消耗的资源非常多,因为它们包含与 null（不存在）数据匹配的数据。在某些情况下,这是不可避免的,但是代价可能非常高。Left join 比 inner join 消耗的资源更多,所以如果可以重新编写查询以使得该查询不使用任何 left join,则会得到非常可观的回报。

在实际应用中,经常会对 left join 和 inner join 两种连接方式产生争议,我对这两种连接方式做了个效率评估。

前提条件:对业务系统中 20 多个表进行表连接生成一个查询 SQL,主表数据量 5 万条,其余表数据量不等。

实验中使用的是某公司产品系统中的 SQL 语句,为保护公司代码产权,语句及过程在这里就不做详细说明。结论（每种连接方式查询运行多次,取平均时间得出）如表 6-2 所示。

表 6-2 执行时间统计

连接类型	平均执行总时间/ms	CPU 执行时间/ms
Inner join	10442	7987
Left join	14754	9898

通过此次实验，可以看到自然连接执行 SQL 查询的时间比左连接方式能节省近 1/3 的时间。

2．系统现有左连接转换为自然连接

如图 6-10 所示为产品 Product 表结构。

列名	数据类型	允许空
ID	varchar(50)	☑
Name	varchar(50)	☑
CategoryID	varchar(50)	☐
		☐

图 6-10　Product 表结构

如图 6-11 所示为类别 Category 表结构。

列名	数据类型	允许空
ID	varchar(50)	☑
Code	varchar(50)	☑
Name	varchar(50)	☑
		☐

图 6-11　Category 表结构

其中 Product 表的 CategoryID 与 Category 表的 ID 是主外键关系。默认情况下，使用以下语句：

```
select * from
[Product] p left join Category c on c.Code = p.CategoryID
```

能过滤出 Product 表中 CategoryID 为 null 或对应类别不存在的数据。但使用:

```
select * from
[Product] p inner join Category c on c.Code = p.CategoryID
```

则过滤不出那些数据。

为解决此问题，可以将自然连接替换为左连接，使 SQL 语句达到相同的功能，避免漏掉 Product 表数据。具体步骤如下：

（1）为 Category 添加“未分类”类别。如图 6-12 所示。

ID	Code	Name
1	A	类别A
2	B	类别B
3	C	类别C
4	NO	未分类
NULL	NULL	NULL

图 6-12　增加“未分类”类别

在表结构不变的情况下，需要为表加一条记录。

（2）通过设置 Product 表的表结构，把 CategoryID 设置为不能为 null。对还未分类的产品可以设置 CategoryID 的值为未分类“NO”类别。如图 6-13 所示。

TSOFT\■SSQL...dbo.Product　TSOFT\MSSQL...dbo.Cate

	ID	Name	CategoryID
	001	产品1	A
	002	产品2	B
	003	产品3	C
	004	产品4	NO
▶	005	产品5	NO
*	NULL	NULL	NULL

图 6-13　应用“未分类”类别

经过以上两个步骤的设置后，使用 inner join 和 left join 两种写法查询出的数据结果是相同的。保证连接的两个表中，左表显示所有数据，右表仅显示匹配的数据，左连接即是如此。

6.3　优化方法指令

由于性能分析指令太多，不能对每一个性能查询指令都像案例那样详细介绍，但一些关键的查询指令我给出了非常详细的注释和说明。本章可以作为性能分析查询字典使用，遇到问题时可以在本章找到对应的监控语句。

使用方法：基本上所有查询指令都是直接在查询分析器里面执行。还要注意一点，如果重启了数据库服务器，那么大部分 DMV 性能查询指令的结果都会丢失，因此使用 DMV 解决性能问题之前先不要重启数据库服务器。

6.3.1　显示查询计划

（1）文本

1）运算符

```
SET SHOWPLAN_TEXT ON
GO
//SQL
SET SHOWPLAN_TEXT OFF
GO
```

2）运算符和估计成本

```
SET SHOWPLAN_ALL ON
GO
//SQL
SET SHOWPLAN_ALL OFF
GO
```

3）运行时信息
SET STATISTICS PROFILE ON
GO
--SQL
SET STATISTICS PROFILE OFF
GO

（2）.XML

1）运算符和估计成本
SET SHOWPLAN_XML ON
GO
--SQL
SET SHOWPLAN_XML OFF
GO

2）运行时信息
SET STATISTICS XML ON
GO
--SQL
SET STATISTICS XML OFF
GO

（3）图形

在查询分析器中，按〈Ctrl+L〉组合键显示预估执行计划，按〈Ctrl+M〉组合键会显示实际执行计划。

6.3.2 查看 SQL 内部执行计划生成/优化信息

```
select * from sys.dm_exec_query_optimizer_info    --如果执行计划已经缓存，则不会输出
DBCC FREEPROCCACHE    --清空编译的执行计划缓存
```

6.3.3 查看缓存对象（syscachobjects）

```
//SQL 2000/2005 及以上版本
SELECT * FROM sys.syscacheobjects;
//SQL 2005 及以上版本
SELECT * FROM sys.dm_exec_cached_plans;    --已缓存的查询计划信息
SELECT * FROM sys.dm_exec_plan_attributes(<handle>);    --接收一个计划句柄作为输入，为计划的每个属性返回一行
SELECT * FROM sys.dm_exec_sql_text(<handle>); --接受一个查询句柄作为输入，返回查询文本
```

6.3.4 清空缓存

```
//Clearing data from cache
DBCC DROPCLEANBUFFERS;
```

```
-- Clearing execution plans from cache:
DBCC FREEPROCCACHE;

-- Clearin execution plans for a particular database:
DBCC FLUSHPROCINDB(15);
```

6.3.5 STATISTICS IO

```
-- First clear cache
    DBCC DROPCLEANBUFFERS;

-- Then run
    SET STATISTICS IO ON;

    SELECT order_id, cust_id, emp_id, orderdate,
    FROM dbo.Orders
    WHERE orderdate >= '20150101'
        AND orderdate < '20180201';
    GO

    SET STATISTICS IO OFF;
    GO
```

6.3.6 STATISTICS TIME

```
-- First clear cache
    DBCC DROPCLEANBUFFERS;
    DBCC FREEPROCCACHE;

-- Then run
    SET STATISTICS TIME ON;

    SELECT order_id, cust_id, emp_id, orderdate
    FROM dbo.Orders
    WHERE orderdate >= '20150101'
        AND orderdate < '20180201';

    SET STATISTICS TIME OFF;
    GO
```

6.3.7 分析执行计划

```
--Example
    SELECT cust_id, emp_id, COUNT(*) AS numorders
    FROM dbo.Orders
```

```
        WHERE orderdate >= '20150201'
            AND orderdate < '20180301'
        GROUP BY cust_id, emp_id
        WITH CUBE;
        GO
```

XML 文件字段解释：

<StmtSimple>：包含 StatementText，SQL 文本。

<StatementSetOptions>：包含多种 SET 选项。

<QueryPlan>具有如下特性：

DegreeOfParallelism：并行计划中每个运算符所拥有的线程数。0 个或 1 个线程数说明这是一个串行计划。

MemoryGrant：在 2-Kbyte 单元中执行这个计划时所授予的内存总量(Kbyte=2×2-Kbyte)。

CachedPlanSize：这个查询计划所消耗的高速缓存总量（Kbyte）。

CompileTime 和 CompileCPU：编译此计划所使用的消耗时间和 CPU 时间（ms）。

CompileMemory：编译此查询时所使用的内存总量（Kbyte）。

<QueryPlan>中还包括<ParameterList>。<ParameterList>中包含每一个参数和变量的编译时间及运行时间值。

<RelOp>包含了一个<MemoryFractions>元素，指出了相应的运算符所使用的总内存授权部分。"输入 faction"是指运算符在读取输入行时所使用的内存授权部分；"输出 faction"是指运算符在生成输出行时所使用的内存授权部分。

1. 取得缓存中每个计划的 XML 计划

```
USE master
GO
CREATE VIEW sp_cacheobjects
(bucketid, cacheobjtype, objtype, objid, dbid, dbidexec, uid,
    refcounts, usecounts, pagesused, setopts, langid, dateformat,
    status, lasttime, maxexectime, avgexectime, lastreads,
    lastwrites, sqlbytes, sql)
AS
SELECT pvt.bucketid,
    CONVERT(nvarchar(17), pvt.cacheobjtype) AS cacheobjtype,
    pvt.objtype,
    CONVERT(int, pvt.objectid) AS object_id,
    CONVERT(smallint, pvt.dbid) AS dbid,
    CONVERT(smallint, pvt.dbid_execute) AS execute_dbid,
    CONVERT(smallint, pvt.user_id) AS user_id,
    pvt.refcounts, pvt.usecounts,
    pvt.size_in_bytes / 8192 AS size_in_bytes,
    CONVERT(int, pvt.set_options) AS setopts,
    CONVERT(smallint, pvt.language_id) AS langid,
    CONVERT(smallint, pvt.date_format) AS date_format,
    CONVERT(int, pvt.status) AS status,
```

```
                CONVERT(bigint, 0),
                CONVERT(bigint, 0),
                CONVERT(bigint, 0),
                CONVERT(bigint, 0),
                CONVERT(bigint, 0),
                CONVERT(int, LEN(CONVERT(nvarchar(max), fgs.text)) * 2),
                CONVERT(nvarchar(3900), fgs.text)
       FROM (SELECT ecp.*, epa.attribute, epa.value
             FROM sys.dm_exec_cached_plans ecp
         OUTER APPLY
                sys.dm_exec_plan_attributes(ecp.plan_handle) epa) AS ecpa
             PIVOT (MAX(ecpa.value) for ecpa.attribute IN
                     (set_options? objectid? dbid?
                         dbid_execute? user_id? language_id?
                         date_format? status?) AS pvt
                OUTER APPLY sys.dm_exec_sql_text(pvt.plan_handle) fgs;
       --sys.syscacheobjects 视图中的所有列
```

2. 过多的重新编译

```
select top 25
    sql_text.text,
    sql_handle,
    plan_generation_num,
    execution_count,
    dbid,
    objectid
from
    sys.dm_exec_query_stats a
    cross apply sys.dm_exec_sql_text(sql_handle) as sql_text
where
    plan_generation_num >1
order by plan_generation_num desc
```

3. 低效的查询计划

```
select
    highest_cpu_queries.plan_handle,
    highest_cpu_queries.total_worker_time,
    q.dbid,
    q.objectid,
    q.number,
    q.encrypted,
    q.[text]
from
    (select top 50
        qs.plan_handle,
        qs.total_worker_time
```

```
            from
                sys.dm_exec_query_stats qs
            order by qs.total_worker_time desc) as highest_cpu_queries
            cross apply sys.dm_exec_sql_text(plan_handle) as q
order by highest_cpu_queries.total_worker_time desc
```

4. 执行计划重用次数

```
select total_elapsed_time / execution_count 平均时间,total_logical_reads/execution_count 逻辑读,
usecounts 重用次数,SUBSTRING(d.text, (statement_start_offset/2) + 1,
        ((CASE statement_end_offset
            WHEN -1 THEN DATALENGTH(text)
            ELSE statement_end_offset END
            - statement_start_offset)/2) + 1) 语句执行  from sys.dm_exec_cached_plans a
cross apply sys.dm_exec_query_plan(a.plan_handle) c
,sys.dm_exec_query_stats b
cross apply sys.dm_exec_sql_text(b.sql_handle) d
where a.plan_handle=b.plan_handle and total_logical_reads/execution_count>4000
ORDER BY total_elapsed_time / execution_count DESC;
```

6.3.8 索引优化

1. 索引优化概要和分析

```
-- 低级 I/O，锁定，闩锁，访问方法信息
SELECT *
FROM sys.dm_db_index_operational_stats(
DB_ID('Performance'), null, null, null);   --数据库 ID，对象 ID，索引 ID，分区 ID

-- 查看不同类型索引的操作计数
SELECT *
FROM sys.dm_db_index_usage_stats;
GO
```

参考：

sys.dm_db_index_usage_stats 收集有关 SQL Server 实例上现有索引的索引使用信息；

sys.dm_exec_query_stats 返回有关缓存查询计划的聚合性能统计信息。它可捕获诸如缓存计划已执行的时间、缓存计划进行时所执行的逻辑和物理读取操作数等信息。

2. 查看索引碎片信息

```
-- 查看索引碎片信息(SQL2005 及以上版本)
-- avg_fragmentation_in_percent 和 avg_page_space_used_in_percent
SELECT *
FROM sys.dm_db_index_physical_stats(
    DB_ID('Performance'), NULL, NULL, NULL, NULL);
GO
```

```sql
--查看指定表的索引情况
select t.name,i.name,s.avg_fragmentation_in_percent from sys.tables t
join sys.indexes i on i.object_id=t.object_id
join sys.dm_db_index_physical_stats(db_id(),object_id('t2'),null,null,'limited') s
on s.object_id=i.object_id and s.index_id=i.index_id

--检查（SQL Server 2005 及以上版本）的索引密度和碎片信息
SELECT i.name                              AS indexname,
       o.name                              AS tablename,
       s.name                              AS schemaname,
       f.index_type_desc                   AS indextype,
       f.avg_page_space_used_in_percent    AS indexdensity,
       f.avg_fragmentation_in_percent      AS indexfragmentation,
       f.page_count                        AS pages
FROM sys.dm_db_index_physical_stats(DB_ID(), NULL, NULL, NULL, 'SAMPLED') f
    INNER JOIN sys.objects o ON o.object_id = f.object_id
    INNER JOIN sys.schemas s ON o.schema_id = s.schema_id
    INNER JOIN sys.indexes i ON f.object_id = i.object_id AND f.index_id = i.index_id
WHERE page_count > 50
    AND f.index_id > 0
ORDER BY o.name, i.index_id

-- 查看索引碎片信息(SQL2000)
-- LogicalFragmentation 和 AveragePageDensity
DBCC SHOWCONTIG WITH ALL_INDEXES, TABLERESULTS, NO_INFOMSGS;
```

参考：

为了获得最佳性能，avg_fragmentation_in_percent 的值应尽可能接近零。但是，0～10％范围内的值都可以接受。所有减少碎片的方法（例如重新生成、重新组织或重新创建）都可降低这个值。

（1）评估磁盘空间使用状况

avg_page_space_used_in_percent 列指示页填充度。为了使磁盘使用状况达到最优，对于没有很多随机插入的索引，此值应接近 100％。但是，对于具有很多随机插入且页很满的索引，其页拆分数将不断增加，从而导致更多的碎片。因此，为了减少页拆分数，此值应小于 100％。使用指定的 FILLFACTOR 选项重新生成索引可以改变页填充度，以便符合索引中的查询模式。此外，ALTER INDEX REORGANIZE 还试图通过将页填充到上一次指定的 FILLFACTOR 来压缩索引，这会增加 avg_space_used_in_percent 的值。注意，ALTER INDEX REORGANIZE 不会降低页填充度。相反，必须执行索引重新生成。

（2）评估索引碎片

碎片由分配单元中同一文件内物理连续的叶级页组成，一个索引至少有一个碎片。索引可以包含的最大碎片数等于索引的页级别页数。碎片越大，意味着读取相同页数所需的磁盘 I/O 越少。所以，avg_fragment_size_in_pages 值越大，范围扫描的性能越好。avg_fragment_size_in_pages 和 avg_fragmentation_in_percent 值成反比，因此，重新生成或重新组织索引会减少碎片数量，但同时会增大碎片大小。

```sql
--查询数据库中所有表的索引密度和碎片信息：
SELECT i.name                              AS indexname,
       o.name                              AS tablename,
       s.name                              AS schemaname,
       f.index_type_desc                   AS indextype,
       f.avg_page_space_used_in_percent    AS indexdensity,
       f.avg_fragmentation_in_percent      AS indexfragmentation,
       f.page_count                        AS pages
FROM sys.dm_db_index_physical_stats(DB_ID(), NULL, NULL, NULL, 'SAMPLED') f
     INNER JOIN sys.objects o ON o.object_id = f.object_id
     INNER JOIN sys.schemas s ON o.schema_id = s.schema_id
     INNER JOIN sys.indexes i ON f.object_id = i.object_id AND f.index_id = i.index_id
WHERE page_count > 50 AND f.index_id > 0 ORDER BY o.name, i.index_id
```

说明："LIMITED""SAMPLED""DETAILED""NULL""DEFAULT"这些模式影响了碎片数据的收集方法。LIMITED 模式扫描堆所有的页，但对于索引，则只扫描叶级上面的父级别页。SAMPLED 收集在堆或索引中 1%采样率的数据。DETAILED 模式扫描所有页（堆或索引）。DETAILED 是执行最慢的，但也是最精确的选项。指定 NULL 或 DEFAULT 的效果与 LIMITED 模式的效果相同。

3. 重新生成和重新组织索引

```sql
-- 重新生成索引[在线]   (SQL2005 及以上版本)
ALTER INDEX idx_cl_od ON dbo.Orders REBUILD WITH (ONLINE = ON);
GO
--为表重新创建索引
ALTER INDEX ALL ON TableName        --TableName
REBUILD WITH (FILLFACTOR = 80, SORT_IN_TEMPDB = ON,
              STATISTICS_NORECOMPUTE = ON);
GO


-- 重新组织索引 (SQL2005 及以上版本)
ALTER INDEX idx_cl_od ON dbo.Orders REORGANIZE;
GO
-- 重新组织索引 (SQL2000)
DBCC INDEXDEFRAG
```

4. 查看没有使用的索引

```sql
Method  1:
DECLARE @dbid INT
SELECT @dbid = DB_ID ( DB_NAME ())
SELECT OBJECTNAME = OBJECT_NAME ( I. OBJECT_ID ),
INDEXNAME = I.NAME ,
I.INDEX_ID
FROM SYS.INDEXES I
```

```
JOIN SYS.OBJECTS O
ON I. OBJECT_ID = O. OBJECT_ID
WHERE OBJECTPROPERTY ( O. OBJECT_ID , 'IsUserTable' ) = 1
AND I.INDEX_ID NOT IN (
SELECT S.INDEX_ID
FROM SYS.DM_DB_INDEX_USAGE_STATS S
WHERE S. OBJECT_ID = I. OBJECT_ID
AND I.INDEX_ID = S.INDEX_ID
AND DATABASE_ID = @dbid )
ORDER BY OBJECTNAME ,
I.INDEX_ID ,
INDEXNAME ASC
GO


Method   2:
SELECT DB_NAME () AS DATABASENAME ,
OBJECT_NAME ( B. OBJECT_ID ) AS TABLENAME ,
B.NAME AS INDEXNAME ,
B.INDEX_ID
FROM SYS.OBJECTS A
INNER JOIN SYS.INDEXES B
ON A. OBJECT_ID = B. OBJECT_ID
WHERE NOT EXISTS ( SELECT *
FROM SYS.DM_DB_INDEX_USAGE_STATS C
WHERE B. OBJECT_ID = C. OBJECT_ID
AND B.INDEX_ID = C.INDEX_ID )
AND A.TYPE <> 'S'
ORDER BY 1 , 2 , 3
```

5. 查看已经使用的索引列表和如何使用索引列表

```
SELECT DB_NAME ( DATABASE_ID ) AS DATABASENAME ,
OBJECT_NAME ( B. OBJECT_ID ) AS TABLENAME ,
INDEX_NAME = ( SELECT NAME
FROM SYS.INDEXES A
WHERE A. OBJECT_ID = B. OBJECT_ID
AND A.INDEX_ID = B.INDEX_ID ),
USER_SEEKS ,
USER_SCANS ,
USER_LOOKUPS ,
USER_UPDATES
FROM SYS.DM_DB_INDEX_USAGE_STATS B
INNER JOIN SYS.OBJECTS C
ON B. OBJECT_ID = C. OBJECT_ID
WHERE DATABASE_ID = DB_ID ( DB_NAME ())
```

```
AND C.TYPE <> 'S'
```

--查看索引统计信息
```
SELECT PVT.TABLENAME , PVT.INDEXNAME , [1] AS COL1 , [2] AS COL2 , [3] AS COL3 ,
[4] AS COL4 , [5] AS COL5 , [6] AS COL6 , [7] AS COL7 , B.USER_SEEKS ,
B.USER_SCANS , B.USER_LOOKUPS
FROM ( SELECT A.NAME AS TABLENAME ,
A. OBJECT_ID ,
B.NAME AS INDEXNAME ,
B.INDEX_ID ,
D.NAME AS COLUMNNAME ,
C.KEY_ORDINAL
FROM SYS.OBJECTS A
INNER JOIN SYS.INDEXES B
ON A. OBJECT_ID = B. OBJECT_ID
INNER JOIN SYS.INDEX_COLUMNS C
ON B. OBJECT_ID = C. OBJECT_ID
AND B.INDEX_ID = C.INDEX_ID
INNER JOIN SYS.COLUMNS D
ON C. OBJECT_ID = D. OBJECT_ID
AND C.COLUMN_ID = D.COLUMN_ID
WHERE A.TYPE <> 'S' ) P
PIVOT
( MIN ( COLUMNNAME )
FOR KEY_ORDINAL IN ( [1] , [2] , [3] , [4] , [5] , [6] , [7] ) ) AS PVT
INNER JOIN SYS.DM_DB_INDEX_USAGE_STATS B
ON PVT. OBJECT_ID = B. OBJECT_ID
AND PVT.INDEX_ID = B.INDEX_ID
AND B.DATABASE_ID = DB_ID ()
ORDER BY TABLENAME , INDEXNAME ;
```

6. 查看每个表的索引结构

```
SELECT A.NAME ,
B.NAME ,
C.KEY_ORDINAL ,
D.NAME
FROM SYS.OBJECTS A
INNER JOIN SYS.INDEXES B
ON A. OBJECT_ID = B. OBJECT_ID
INNER JOIN SYS.INDEX_COLUMNS C
ON B. OBJECT_ID = C. OBJECT_ID
AND B.INDEX_ID = C.INDEX_ID
INNER JOIN SYS.COLUMNS D
ON C. OBJECT_ID = D. OBJECT_ID
AND C.COLUMN_ID = D.COLUMN_ID
```

WHERE A.TYPE <> 'S'
ORDER BY 1 , 2 , 3

7. 查看索引列顺序

SELECT TABLENAME , INDEXNAME , INDEXID , [1] AS COL1 , [2] AS COL2 , [3] AS COL3 ,
[4] AS COL4 , [5] AS COL5 , [6] AS COL6 , [7] AS COL7
FROM (SELECT A.NAME AS TABLENAME ,
B.NAME AS INDEXNAME ,
B.INDEX_ID AS INDEXID ,
D.NAME AS COLUMNNAME ,
C.KEY_ORDINAL
FROM SYS.OBJECTS A
INNER JOIN SYS.INDEXES B
ON A. OBJECT_ID = B. OBJECT_ID
INNER JOIN SYS.INDEX_COLUMNS C
ON B. OBJECT_ID = C. OBJECT_ID
AND B.INDEX_ID = C.INDEX_ID
INNER JOIN SYS.COLUMNS D
ON C. OBJECT_ID = D. OBJECT_ID
AND C.COLUMN_ID = D.COLUMN_ID
WHERE A.TYPE <> 'S') P
PIVOT
(MIN (COLUMNNAME)
FOR KEY_ORDINAL IN ([1] , [2] , [3] , [4] , [5] , [6] , [7])) AS PVT
--where tablename = 'AA_Currency'
ORDER BY TABLENAME , INDEXNAME ;

8. 同时查看已使用和未使用索引及列位置信息

SELECT PVT.TABLENAME , PVT.INDEXNAME , PVT.INDEX_ID , [1] AS COL1 , [2] AS COL2 , [3] AS COL3 ,
[4] AS COL4 , [5] AS COL5 , [6] AS COL6 , [7] AS COL7 , B.USER_SEEKS ,
B.USER_SCANS , B.USER_LOOKUPS
FROM (SELECT A.NAME AS TABLENAME ,
A. OBJECT_ID ,
B.NAME AS INDEXNAME ,
B.INDEX_ID ,
D.NAME AS COLUMNNAME ,
C.KEY_ORDINAL
FROM SYS.OBJECTS A
INNER JOIN SYS.INDEXES B
ON A. OBJECT_ID = B. OBJECT_ID
INNER JOIN SYS.INDEX_COLUMNS C
ON B. OBJECT_ID = C. OBJECT_ID
AND B.INDEX_ID = C.INDEX_ID
INNER JOIN SYS.COLUMNS D
ON C. OBJECT_ID = D. OBJECT_ID

```
         AND C.COLUMN_ID = D.COLUMN_ID
     WHERE A.TYPE <> 'S' ) P
 PIVOT
 ( MIN ( COLUMNNAME )
 FOR KEY_ORDINAL IN ( [1] , [2] , [3] , [4] , [5] , [6] , [7] ) ) AS PVT
 INNER JOIN SYS.DM_DB_INDEX_USAGE_STATS B
 ON PVT. OBJECT_ID = B. OBJECT_ID
 AND PVT.INDEX_ID = B.INDEX_ID
 AND B.DATABASE_ID = DB_ID ()
 UNION
 SELECT TABLENAME , INDEXNAME , INDEX_ID , [1] AS COL1 , [2] AS COL2 , [3] AS COL3 ,
 [4] AS COL4 , [5] AS COL5 , [6] AS COL6 , [7] AS COL7 , 0 , 0 , 0
 FROM ( SELECT A.NAME AS TABLENAME ,
     A. OBJECT_ID ,
     B.NAME AS INDEXNAME ,
     B.INDEX_ID ,
     D.NAME AS COLUMNNAME ,
     C.KEY_ORDINAL
     FROM SYS.OBJECTS A
     INNER JOIN SYS.INDEXES B
     ON A. OBJECT_ID = B. OBJECT_ID
     INNER JOIN SYS.INDEX_COLUMNS C
     ON B. OBJECT_ID = C. OBJECT_ID
     AND B.INDEX_ID = C.INDEX_ID
     INNER JOIN SYS.COLUMNS D
     ON C. OBJECT_ID = D. OBJECT_ID
     AND C.COLUMN_ID = D.COLUMN_ID
     WHERE A.TYPE <> 'S' ) P
 PIVOT
 ( MIN ( COLUMNNAME )
 FOR KEY_ORDINAL IN ( [1] , [2] , [3] , [4] , [5] , [6] , [7] ) ) AS PVT
 WHERE NOT EXISTS ( SELECT OBJECT_ID ,
 INDEX_ID
 FROM SYS.DM_DB_INDEX_USAGE_STATS B
 WHERE DATABASE_ID = DB_ID ( DB_NAME ())
 AND PVT. OBJECT_ID = B. OBJECT_ID
 AND PVT.INDEX_ID = B.INDEX_ID )
 ORDER BY TABLENAME , INDEX_ID ;
```

9. 查看聚集表个数

```
declare    @tableName    varchar(200)
declare @count int

select   @count = 0
declare    tb_cursor    cursor       for
```

```
select name from sysobjects where xtype ='U' and name <> 'dtproperties'
open    tb_cursor
fetch   next    from    tb_cursor   into    @tableName;
while   @@fetch_status = 0
begin
if (objectproperty(object_id(@tableName),'TableHasClustIndex') = 1)
begin
print @tableName
set @count = @count + 1
end
fetch   next    from    tb_cursor   into    @tableName;
end
close   tb_cursor
deallocate  tb_cursor
print '聚集表个数：' + cast( @count   as    varchar)
```

10. 索引使用效率评估

```
-- 使用很少的索引排在最先
declare @dbid int
select @dbid = db_id()
select objectname=object_name(s.object_id), s.object_id, indexname=i.name, i.index_id
        , user_seeks, user_scans, user_lookups, user_updates
from sys.dm_db_index_usage_stats s,
        sys.indexes i
where database_id = @dbid and objectproperty(s.object_id,'IsUserTable') = 1
and i.object_id = s.object_id
and i.index_id = s.index_id
order by (user_seeks + user_scans + user_lookups + user_updates) asc
```

6.3.9 数据库和文件空间

1. 查看数据库空间

```
-- sp_spaceused：查看数据库空间

SELECT DB_NAME() AS DbName,name AS FileName,size/128.0 AS CurrentSizeMB,
size/128.0 - CAST(FILEPROPERTY(name, 'SpaceUsed') AS INT)/128.0 AS FreeSpaceMB
FROM sys.database_files
```

2. 查看数据库中每个表的占用空间

```
--这个有点问题，导致表重复
select  object_name(id)   tablename,8*reserved/1024    reserved,rtrim(8*dpages/1024)+'Mb'    used,8*
(reserved-dpages)/1024 unused,8*dpages/1024-rows/1024*minlen/1024 free,
rows
from sysindexes
--where indid=1
```

order by reserved desc

--包含索引和数据 Size
Drop table #temp
Create table #temp (Name nvarchar(255), Rows int, Reserved nvarchar(50), Data nvarchar(50), Index_size nvarchar(50), Unused nvarchar(50));
exec sp_MSforeachtable 'insert into #temp exec sp_spaceused "?"';
select * from #temp where name = 'ST_RDRecord_b'

--显示表的分布情况
DBCC SHOWCONTIG(IndexFragment_Bad)

查看表空间情况：

```sql
if not exists (select * from dbo.sysobjects where id = object_id(N'[dbo].[tablespaceinfo]') and OBJECTPROPERTY(id, N'IsUserTable') = 1)
    create table   tablespaceinfo                          --创建结果存储表
                (nameinfo varchar(50) ,
                 rowsinfo int , reserved varchar(20) ,
                 datainfo varchar(20)   ,
                 index_size varchar(20) ,
                 unused varchar(20) )

delete from tablespaceinfo --清空数据表
declare @tablename varchar(255)     --表名称
declare @cmdsql varchar(500)
DECLARE Info_cursor CURSOR FOR
select o.name
from dbo.sysobjects o where OBJECTPROPERTY(o.id, N'IsTable') = 1
    and o.name not like N'#%%'    order by o.name
OPEN Info_cursor
FETCH NEXT FROM Info_cursor
INTO @tablename
WHILE @@FETCH_STATUS = 0
BEGIN
    if exists (select * from dbo.sysobjects where id = object_id(@tablename) and OBJECTPROPERTY(id, N'IsUserTable') = 1)
        execute sp_executesql
            N'insert into tablespaceinfo   exec sp_spaceused @tbname',
            N'@tbname varchar(255)',
            @tbname = @tablename
    FETCH NEXT FROM Info_cursor
    INTO @tablename
END
CLOSE Info_cursor
DEALLOCATE Info_cursor
GO
```

```
//knowsky.com 数据库信息
sp_spaceused @updateusage = 'TRUE'
//表信息
select *
from tablespaceinfo
order by cast(left(ltrim(rtrim(reserved)) , len(ltrim(rtrim(reserved)))-2) as int) desc
```

查询结果字段的解释如表 6-3 所示。

表 6-3 查询结果字段解释

Name	nvarchar(20)	为其请求空间使用信息的表名
Rows	char(11)	表中现有的行数
Reserved	varchar(18)	表保留的空间总量
Data	varchar(18)	表中的数据所使用的空间量
index_size	varchar(18)	表中的索引所使用的空间量
Unused	varchar(18)	表中未用的空间量

3. 查看数据库页面信息

```
DBCC   TRACEON(3604)
GO
DBCC EXTENTINFO(Performance)
GO
```

4. 获取数据文件头部信息

```
DBCC TRACEON(3604)
GO
DBCC PAGE(Performance,1,0,3) – 1: 文件 ID, 从 15 节获得
GO


//内存缓冲区中的头部信息*/BUFFER:
//内存缓冲区中的头部信息*/BUFFER:
//页面头部信息*/PAGE HEADER:
//数据文件主要管理页面分配信息*/Allocation Status:
//数据文件头部信息*/File Header Data:
//数据文件头部存储的数据转换后的值*/BindingID = 234c3f4c-53ac-45f2-8329-615c2f5ddaba
```

数据文件的头部信息如表 6-4 所示。

表 6-4 数据文件头部信息字段解释

字 段 名 称	说　　明
BindingID	给数据文件产生的全局唯一标识符
FileGroupId	文件组的 ID

(续)

字段名称	说明
FileIdProp	在数据库中的文件 ID
Size	当前空间大小，以页面为单位
MaxSize	最大空间大小，65535 表示无限大小
Growth	增长值
Perf	保留
BackupLsn	数据页面中上一个产生日志记录备份时的 LSN
MaxLsn	最大的日志序列号
FirstLsn	首次日志序列号
FirstCreateIndexLsn	首次创建索引时的日志序列号
FirstUpdateLsn	首次更新时的日志序列号
FirstNonloggedUpdateLsn	首次非日志记录时的日志序列号
CreateLsn	创建时的日志序列号
DifferentialBaseLsn	发生差异更改时的日志序列号
DifferentialBaseGuid	发生差异更改时的全局唯一标识符
MinSize	最小的大小
Status	文件状态
UserShrinkSize	用户收缩文件的大小

5. 查看数据库文件表信息

select * from sys.master_files
go

6. 查看数据库信息

sp_helpdb Performance

7. 查看数据库日志文件信息

DBCC SQLPERF(logspace)

8. 查看表中有多少重定向的行

SELECT forwarded_record_count
FROM sys.dm_db_index_physical_stats
 (db_id('Northwind2'), object_id('TableName'),
  null, null, 'detailed');

9. 查看数据库的版本和补丁信息

select
CONVERT(sysname, SERVERPROPERTY('servername')) N'数据库实例名',

SERVERPROPERTY('ProductVersion') N'数据库版本',
SERVERPROPERTY ('Edition') N'数据库类型',
SERVERPROPERTY('ProductLevel') N'数据库补丁'

10. 查看表占用空间情况

dbcc showcontig('table_name')

11. 查看数据库文件

select * from sysfiles

select   文件标识号=fileid,
cast(size*8/1024.   as   decimal(10,2))   [文件大小(MB)],
name   逻辑文件名
filename   物理文件名
from   sysfiles

6.3.10 监视命令

1. 用户和进程的快照信息

sp_who 报告有关当前 SQL Server 用户和进程的快照信息,包括当前正在执行的语句以及该语句是否被阻塞。

(1) 查看当前连接:

SQL General Statistics: User Connection
SQL
SELECT * FROM
    [Master].[dbo].[SYSPROCESSES] WHERE [DBID] IN (SELECT
        [DBID]
    FROM
        [Master].[dbo].[SYSDATABASES]
    WHERE
        NAME='ldht_xzysb'   --DBName
)
sp_who 'sa'

(2) 查询超过 10min 没有做任何动作的连接

select   *   from   master.dbo.sysprocesses
where   spid>50
and   waittype   =   0x0000
and   waittime   =   0
and   status   =   'sleeping'
and   last_batch   <   dateadd(minute,   -10,   getdate())
and   login_time   <   dateadd(minute,   -10,   getdate())

2. 报告有关锁的快照信息

sp_lock 报告有关锁的快照信息，包括对象 ID、索引 ID、锁的类型以及锁应用于的类型或资源。

3. 显示统计信息(CPU/IO 等)

sp_monitor 显示统计信息，包括 CPU 使用率、I/O 使用率以及自上次执行 sp_monitor 以来的空闲时间。

--显示索引的统计信息
DBCC SHOW_STATISTICS('Orders','cust_date_indx')

update statistics TableName;   --更新统计信息

4. 内置函数

内置函数可显示自启动服务器以来有关 SQL Server 活动的快照统计信息，这些统计信息存储在预定义的 SQL Server 计数器中。例如，@@CPU_BUSY 包含 CPU 执行 SQL Server 代码所持续的时间；@@CONNECTIONS 包含 SQL Server 连接或尝试连接的次数；@@PACKET_ERRORS 包含 SQL Server 连接上出现的网络数据包数。

5. 数据库使用内存情况

select physical_memory_in_bytes,virtual_memory_in_bytes,* from sys.dm_os_sys_info

6.3.11 SQL 性能统计

1. 语句当前正在使用的资源，以及需要对其进行检查以改进性能的语句

```
SELECT
    substring(text,qs.statement_start_offset/2
        ,(CASE
            WHEN qs.statement_end_offset = -1 THEN len(convert(nvarchar(max), text)) * 2
            ELSE qs.statement_end_offset
        END - qs.statement_start_offset)/2)
    ,qs.plan_generation_num as recompiles
    ,qs.execution_count as execution_count
    ,qs.total_elapsed_time - qs.total_worker_time as total_wait_time
    ,qs.total_worker_time as cpu_time
    ,qs.total_logical_reads as reads
    ,qs.total_logical_writes as writes
FROM sys.dm_exec_query_stats qs
    CROSS APPLY sys.dm_exec_sql_text(qs.sql_handle) st
    LEFT JOIN sys.dm_exec_requests r
        ON qs.sql_handle = r.sql_handle
ORDER BY 3 DESC
```

2. 执行 SQL 次数和逻辑次数

SELECT   creation_time   N'语句编译时间'

```
    ,last_execution_time    N'上次执行时间'
    ,total_physical_reads N'物理读取总次数'
    ,total_logical_reads/execution_count N'每次逻辑读取次数'
    ,total_logical_reads    N'逻辑读取总次数'
    ,total_logical_writes N'逻辑写入总次数'
    , execution_count    N'执行次数'
    , total_worker_time/1000 N'所用的 CPU 总时间 ms'
    , total_elapsed_time/1000    N'总花费时间 ms'
    , (total_elapsed_time / execution_count)/1000    N'平均时间 ms'
    ,SUBSTRING(st.text, (qs.statement_start_offset/2) + 1,
        ((CASE statement_end_offset
          WHEN -1 THEN DATALENGTH(st.text)
          ELSE qs.statement_end_offset END
          - qs.statement_start_offset)/2) + 1) N'执行语句'
FROM sys.dm_exec_query_stats AS qs
CROSS APPLY sys.dm_exec_sql_text(qs.sql_handle) st
where SUBSTRING(st.text, (qs.statement_start_offset/2) + 1,
        ((CASE statement_end_offset
          WHEN -1 THEN DATALENGTH(st.text)
          ELSE qs.statement_end_offset END
          - qs.statement_start_offset)/2) + 1) not like '%fetch%'
ORDER BY    total_elapsed_time / execution_count DESC;
```

6.3.12 跟踪文件统计

1．读取跟踪文件中的数据

```
SELECT *
FROM fn_trace_gettable('c:\inside_sql.trc', 1);


SELECT *
INTO inside_sql_trace
FROM fn_trace_gettable('c:\inside_sql.trc', 1);
```

2．查看跟踪表里前 10%的查询

```
SELECT *
FROM
(
  SELECT
     *,
     NTILE(10) OVER(ORDER BY Duration) Bucket
  FROM TraceTable
) x
WHERE Bucket = 10;
```

3. 找出跟踪表里的前几位查询

```sql
SELECT
   QueryForm,
   AVG(Duration),
   AVG(Reads),
   AVG(Writes),
   AVG(CPU)
FROM
(
   SELECT
      dbo.fn_sqlsig(TextData) AS QueryForm,
      1. * Duration AS Duration,
      1. * Reads AS Reads,
      1. * Writes AS Writes,
      1. * CPU AS CPU
   FROM TraceTable
   WHERE TextData IS NOT NULL
) x
GROUP BY QueryForm;
```

4. 查找所有用户异常和断开的连接，查询相关错误及造成的故障

```sql
WITH Exceptions AS
(
   SELECT
      T0.SPID,
      T0.EventSequence,
      COALESCE(T0.TextData, 'Attention') AS Exception,
      T1.TextData AS MessageText
   FROM TraceTable T0
      LEFT OUTER JOIN TraceTable T1 ON
         T1.EventSequence = T0.EventSequence + 1
         AND T1.EventClass = 162
   WHERE
      T0.EventClass IN (16, 33)
      AND (T0.EventClass = 16 OR T1.EventSequence IS NOT NULL)
)
SELECT *
FROM Exceptions
CROSS APPLY
(
   SELECT TOP(1)
      TextData AS QueryText
   FROM TraceTable Queries
   WHERE
      Queries.SPID = Exceptions.SPID
```

```
        AND Queries.EventSequence < Exceptions.EventSequence
        AND Queries.EventClass IN (10, 13)
    ORDER BY EventSequence DESC
) p;
```

6.4 性能故障检测方法

上一节是从整体层面讲了一下 SQL Server 的各种工具及对性能的支持。这一节主要从症状入手，直接切入一些分析问题的方法，比如遇到 CPU 问题、内存问题、I/O 问题、tempdb 问题、阻塞/死锁问题以及存储问题时，可以通过查找这些分析方法进行 SQL 定位。

6.4.1 CPU 问题诊断

CPU 状态分为运行中（RUNNING）、可运行（RUNNABLE）和悬挂（SUSPENED）。如果有大量的工作进程处在可运行状态，CPU 就出现了瓶颈的征兆；如果工作进程在悬挂状态时间较长，则说明 SQL Server 可能有过多的阻塞。

1．可运行状态下的工作进程数量

```sql
SELECT COUNT(*) AS workers_waiting_for_cpu, t2.Scheduler_id
FROM sys.dm_os_workers AS t1, sys.dm_os_schedulers AS t2
WHERE t1.state = 'RUNNABLE' AND
    t1.scheduler_address = t2.scheduler_address AND
    t2.scheduler_id < 255
GROUP BY t2.scheduler_id;
```

2．工作进程在可运行状态下花费的时间

```sql
SELECT SUM(signal_wait_time_ms)
FROM sys.dm_os_wait_stats;
```

3．每次执行过程中占用 CPU 最多的前 10 位

```sql
SELECT TOP 10
    total_worker_time/execution_count AS avg_cpu_cost,
    plan_handle, execution_count,
    (SELECT SUBSTRING(text, statement_start_offset/2 + 1,
        (CASE WHEN statement_end_offset = -1
            THEN LEN(CONVERT(nvarchar(max), text)) * 2
            ELSE statement_end_offset
        END - statement_start_offset)/2)
    FROM sys.dm_exec_sql_text(sql_handle)) AS query_text
FROM sys.dm_exec_query_stats
ORDER BY [avg_cpu_cost] DESC;
```

4．每次执行过程中运行最频繁的前 10 位（与占用 CPU 最多的前 10 位通用）

```sql
SELECT TOP 10 total_worker_time, plan_handle,execution_count,
```

```
        (SELECT SUBSTRING(text, statement_start_offset/2 + 1,
            (CASE WHEN statement_end_offset = -1
                THEN LEN(CONVERT(nvarchar(max),text)) * 2
                ELSE statement_end_offset
            END - statement_start_offset)/2)
        FROM sys.dm_exec_sql_text(sql_handle)) AS query_text
    FROM sys.dm_exec_query_stats
    ORDER BY execution_count DESC;
```

5．编译和重编译

过度编译和重编译是 CPU 密集型活动，发生大量重编译会导致 CPU 利用率增加。用下列监视器查看编译和重编译的速度：

- SQLServer:SQL Statistics: Batch Requests/Sec（统计每秒批处理请求数）
- SQLServer:SQL Statistics: SQL Compilations/Sec（统计每秒 SQL 编译次数）
- SQLServer:SQL Statistics: SQL Recompilations/Sec（统计每秒 SQL 重编译次数）

（1）SQL Server 在优化查询计划上花费的时间

可以用下面的 DMV 查询 SQL Server 在优化查询计划上花费的时间。

```
SELECT *
FROM sys.dm_exec_query_optimizer_info
WHERE counter = 'optimizations' OR counter = 'elapsed time';
```

（2）找出编译得最多的前 10 位查询计划（SQL2005 及以上版本）。

```
SELECT TOP 10 plan_generation_num, execution_count,
        (SELECT SUBSTRING(text, statement_start_offset/2 + 1,
            (CASE WHEN statement_end_offset = -1
                THEN LEN(CONVERT(nvarchar(max),text)) * 2
                ELSE statement_end_offset
            END - statement_start_offset)/2)
        FROM sys.dm_exec_sql_text(sql_handle)) AS query_text
    FROM sys.dm_exec_query_stats
    WHERE plan_generation_num >1
    ORDER BY plan_generation_num DESC;
```

6．获取分配给用于存储优化查询计划过程的高速缓存的内存

使用 DBCC 命令获取分配信息。

```
DBCC MEMORYSTATUS
```

参考：程序调整缓存下的 TotalPages 指的是用来存储优化计划的缓存缓冲池页。通过与基线对比，如果此值缩小，可能暗示 SQL 内存压力增大。

7．确定服务器的活动

```
SELECT es.session_id
    ,es.program_name
    ,es.login_name
```

```
            ,es.nt_user_name
            ,es.login_time
            ,es.host_name
            ,es.cpu_time
            ,es.total_scheduled_time
            ,es.total_elapsed_time
            ,es.memory_usage
            ,es.logical_reads
            ,es.reads
            ,es.writes
            ,st.text
        FROM sys.dm_exec_sessions es
            LEFT JOIN sys.dm_exec_connections ec
                ON es.session_id = ec.session_id
            LEFT JOIN sys.dm_exec_requests er
                ON es.session_id = er.session_id
            OUTER APPLY sys.dm_exec_sql_text (er.sql_handle) st
        WHERE es.session_id > 50      -- < 50 system sessions
        ORDER BY es.cpu_time DESC
```

参考：我发现以上有助于确定需要重点了解哪些应用程序。当将一个应用程序的各个会话的 CPU、内存、读取、写入和逻辑读取进行比较，并确定 CPU 资源要远远高于正在使用的其他资源时，我开始重点关注这些 SQL 语句。

6.4.2 内存诊断

1．物理内存压力的检测

查看任务管理器，物理内存可用数。当数值降低到 50~100MB 之间时，应该开始关注；当该数值小于 10MB 时，就会出现外部内存压力。

以下计数器可以确认内存是否存在压力：

➢ 内存：可用字节。

可用物理内存总量，即清零、空闲及备用的内存列表上的空间总量之和。

➢ SQL Server：缓存管理器：缓存命中率。

指不用通过磁盘读取而直接在缓冲池中找到页的比例，正常应该为 90%左右。

➢ SQL Server：缓存管理器：页平均寿命。

指一个没有被引用的页在缓冲区池中保留的秒数。如果数值比较低，则说明缓冲区池遇到了内存不足的情况。

➢ SQL Server：缓存管理器：检查点页数/秒。

指被检查点刷新的页数，或者要求所有脏页被刷新的其他操作的数目。它能显示工作负荷中增加的缓存区池活动量。

➢ SQL Server：缓存管理器：延迟写入/秒。

指的是缓冲管理器的延迟写入器写入的缓冲数目，它的作用类似于前面提到的检查点页/秒，如图 6-14 所示。

Buffer Counts	Buffers
1 Committed	1920
2 Target	104853
3 Hashed	713
4 Stolen Pote...	191306
5 External Re...	0
6 Min Free	64
7 Visible	104853
8 Available P...	373559

图 6-14　查询结果

> 已提交

此值显示了已提交的所有缓冲和相关联的物理内存，已提交的值是缓冲区的当前大小。如果启用了 AWE 支持，此值还包含分配的物理内存。

> 目标

此值显示了缓冲区的目标大小。SQL Server 降低这个值以响应 Windows 操作系统中的内存下降通知。目标页面数减少表明它正在响应外部物理内存的压力。

物理内存压力可能来自 SQL Server 自身，如果高速缓存增加过快，就会给缓存区池带来压力，这时通过检查缓冲分配中的所有异常来查看 SQL Server 内部的内存分配。可以查看 DBCC MEMORYSTATUS 输出中的窃取页数，如图 6-15 所示。

Buffer Distribution	Buffers
1 Stolen	385
2 Free	81
3 Cached	741
4 Database (clean)	677
5 Database (dirty)	36
6 I/O	0
7 Latched	0

图 6-15　查询结果

如果所有已提交的页面中窃取页的比率很高（高于 75%～80%），则说明有内部的内存压力。

2．虚拟内存压力的检测

系统监视计数器如下：

> 分页文件

%使用率，指的是以百分比表示页面文件实例中正在使用的量。

> 内存：提交极限

指不用扩展分页文件就可以提交的虚拟内存量（以字节计）。已提交内存是在磁盘分页文件中有预留空间的物理内存。

3．内存压力的隔离和排查

系统中的物理内存使用总量可以通过叠加下列计数器得到一个粗略的结果：

> 进程：工作集

指的是单个进程的计数器。
- 内存：高速缓存字节

指的是系统工作集的计数器。
- 内存：未分页字节

指的是未分页池大小的计数器。
- 内存：可用字节

指的是等同于任务管理器中的可用值。

如果不存在物理内存压力，进程的私有字节或任务管理器中的 VM 值应该会接近进程的工作集大小，这就意味着没有调出内存。

（1）使用下面的 DMV 查询来找出缓冲区池消耗内存的问题（包括 AWE）

```
SELECT  SUM(multi_pages_kb + virtual_memory_committed_kb
            + shared_memory_committed_kb
            + awe_allocated_kb) AS [Used by BPool, Kb]
FROM sys.dm_os_memory_clerks
WHERE type = 'MEMORYCLERK_SQLBUFFERPOOL';
```

（2）查询并确认是哪些内部组件窃取了缓冲区池中大部分的页面

如果已经确定（通过 DBCC MEMORYSTATUS）当前内部内存不足的原因在于内部组件从缓冲区池中窃取了绝大多数页面，这时可以用下列 DMV 查询确认是哪些内部组件窃取了缓冲区池中大部分的页面。

```
SELECT TOP 10 type, SUM(single_pages_kb) AS stolen_mem_kb
FROM sys.dm_os_memory_clerks
GROUP BY type
ORDER BY SUM(single_pages_kb) DESC;
```

（3）用包含下列查询的多页分配器来确认在缓冲区池外分配了内存的内部组件

内部组件使用的内存无法控制。然而，定位出使用内存最多的内部组件有助于缩小故障的调查范围。可以用包含下列查询的多页分配器来确认在缓冲区池外分配了内存的内部组件。

```
SELECT type, SUM(multi_pages_kb) AS memory_allocated_KB
FROM sys.dm_os_memory_clerks
WHERE multi_pages_kb != 0
GROUP BY type;
```

4．设置 min/max memory

```
sp_configure 'min server memory', 1024
RECONFIGURE
GO
sp_configure 'max server memory', 6144
RECONFIGURE
GO
```

6.4.3 I/O 诊断

1. I/O 瓶颈的检测

> 物理磁盘对象：磁盘队列平均长度。
> 物理磁盘对象：磁盘每次读/写平均用时（以秒计）。

指的是每次从磁盘中读取与写入数据时的平均用时，通用指标如下：

◆ 小于 10ms，非常好。
◆ 10ms～20ms 之间，还过得去。
◆ 20ms～50ms 之间，很慢，需要多加注意。
◆ 大于 50ms 则被认为是严重的 I/O 瓶颈。

物理磁盘：磁盘每秒读/写数。

指的是磁盘上读/写操作的速度。一定要确保该数字小于磁盘容量的 85%。当超过磁盘容量的 85%时，磁盘访问时间将呈指数级增长。

（1）文件级 DMV 查询

```
SELECT database_id, file_id, io_stall_read_ms, io_stall_write_ms
FROM sys.dm_io_virtual_file_stats(NULL, NULL);
```

执行结果中 io_stall_read_ms 和 io_stall_write_ms 这两列表示的是 SQL Server 启动后向文件发出读和写指令的等待时间。要取得有意义的数据，需要在短时间取得这些数据的快照，然后将它们同基线数据比较。

（2）查询可以用来查找 I/O 等待锁的统计信息

检查等待锁来确认全部的 I/O 瓶颈。当一个页被读/写访问且该页不在缓冲区中时，即引发等待，根据请求类型的不同以 PAGEIOLATCH_EX 或 PAGEIOLATCH_SH 模式等待。以下 DMV 查询可以用来查找 I/O 等待锁的统计信息：

```
SELECT wait_type, waiting_tasks_count, wait_time_ms, signal_wait_time_ms
FROM sys.dm_os_wait_stats
WHERE wait_type LIKE 'PAGEIOLATCH%'
ORDER BY wait_type;
```

显示列表中，最重要的等待锁是 PAGEIOLATCH_EX 或 PAGEIOLATCH_SH。当一个任务在锁存器上等待一个 I/O 请求中的缓冲区时，这两种等待锁就会启动。如果等待时间过长说明磁盘子系统出了故障。Wait_time_ms 列包括一个工作进程在悬挂状态下花费的时间和在可运行状态下花费的时间，而 signal_wait_time_ms 表示的只是一个工作进程在可运行状态下花费的时间，因此两者之差实际上代表等待 I/O 完成所花的时间。要取得有意义的数据，需要查看感兴趣的时间间隔。

（3）I/O 瓶颈

```
select top 25
    (total_logical_reads/execution_count) as avg_logical_reads,
    (total_logical_writes/execution_count) as avg_logical_writes,
    (total_physical_reads/execution_count) as avg_phys_reads,
```

```
        Execution_count,
        statement_start_offset as stmt_start_offset,
        sql_handle,
        plan_handle
    from sys.dm_exec_query_stats
    order by
    (total_logical_reads + total_logical_writes) desc
```

2．I/O 瓶颈的隔离和排查

内存不足也会导致 I/O 瓶颈。例如，如果没有足够的物理内存，缓冲区池中的页面就会不断地回收，引起物理 I/O，最终导致 I/O 瓶颈。

（1）查看引发 I/O 最多的前 10 位的查询或批处理

```
SELECT TOP 10
    (total_logical_reads/execution_count) AS avg_logical_reads,
    (total_logical_writes/execution_count) AS avg_logical_writes,
    (total_physical_reads/execution_count) AS avg_phys_reads,
     execution_count,plan_handle,
    (SELECT SUBSTRING(text, statement_start_offset/2 + 1,
        (CASE WHEN statement_end_offset = -1
            THEN LEN(CONVERT(nvarchar(MAX),text)) * 2
            ELSE statement_end_offset
        END - statement_start_offset)/2)
     FROM sys.dm_exec_sql_text(sql_handle)) AS query_text
FROM sys.dm_exec_query_stats
ORDER BY (total_logical_reads + total_logical_writes) DESC;
```

（2）显示缓存中累计逻辑读取次数最高的 20 个查询的文本和执行计划

```
SELECT TOP 20 SUBSTRING(qt.text, (qs.statement_start_offset/2)+1,
        ((CASE qs.statement_end_offset
            WHEN -1 THEN DATALENGTH(qt.text)
            ELSE qs.statement_end_offset
         END - qs.statement_start_offset)/2)+1),
    qs.execution_count,
    qs.total_logical_reads, qs.last_logical_reads,
    qs.min_logical_reads, qs.max_logical_reads,
    qs.total_elapsed_time, qs.last_elapsed_time,
    qs.min_elapsed_time, qs.max_elapsed_time,
    qs.last_execution_time,
    qp.query_plan
FROM sys.dm_exec_query_stats qs
CROSS APPLY sys.dm_exec_sql_text(qs.sql_handle) qt
CROSS APPLY sys.dm_exec_query_plan(qs.plan_handle) qp
WHERE qt.encrypted=0
ORDER BY qs.total_logical_reads DESC
```

（3）缺失索引也是导致 I/O 过多的原因之一，查看哪些索引已经丢失且可能会被使用

```
SELECT t1.object_id, t2.user_seeks, t2.user_scans,
       t1.equality_columns, t1.inequality_columns
FROM sys.dm_db_missing_index_details AS t1,
     sys.dm_db_missing_index_group_stats AS t2,
     sys.dm_db_missing_index_groups AS t3
WHERE database_id = DB_ID()
    AND object_id = OBJECT_ID('t_sample')    --TableName
    AND t1.index_handle = t3.index_handle
    AND t2.group_handle = t3.index_group_handle;

--显示缺失索引
SELECT * FROM sys.dm_db_missing_index_details
(依赖：SET STATISTICS XML ON)
```

创建索引规则：
- 首先列出相等列（位于列表的最左侧）。
- 在相等列之后列出不等列（位于列出的相等列的右侧）。
- 在 CREATE INDEX 语句的 INCLUDE 子句中列出包含列。
- 若要确定相等列的有效顺序，可根据其进行选择排序，即首先列出最具选择性的列。

6.4.4 tempdb 诊断

1. tempdb 性能问题的检测

（1）查询是否存在一个或多个等待获取 tempdb 中页面锁存器的线程

```
SELECT session_id, wait_duration_ms, resource_description
FROM sys.dm_os_waiting_tasks
WHERE wait_type LIKE 'PAGE%LATCH_%'
    AND resource_description like '2:%';
```

这个 DMV 显示正在等待的进程，要想确认分配瓶颈，需要经常轮询该 DMV。

（2）使用系统监视计数器监视 tempdb 中用户和内部对象分配/回收活动的异常增长

- SQL Server: Access Methods: Worktables Created/Sec

每秒钟所创建的工作表的数量。工作表是临时对象，用来存储缓冲查询、LOB 变量和游标的结果。参考值<200，但仍然要把它同基线相比较。

- SQL Server: Access Methods: Workfiles Created/Sec

每秒钟所创建的工作文件的数量。工作文件和工作表类似，但是工作文件是严格由哈希操作来创建的。工作文件用来存储哈希连接和哈希聚合的临时结果。

- SQL Server: Access Methods: Worktables from Cache Ratio

最初两页未被分配但可从工作表高速缓存中直接获得创建工作表数量的比例。在 SQL 2000 中没有临时表的高速缓存。

- SQL Server: General Statistics: Temp Tables Creation Rate

每秒钟创建的临时表或临时变量的数量。
- SQL Server: General Statistics: Temp Tables for Destruction

等待被清理系统线程销毁的临时表或临时变量的数量。

2. tempdb 瓶颈的隔离和排查

（1）查询 tempdb 中当前引发最多分配和回收操作的执行查询

```
SELECT TOP 10 t1.session_id, t1.request_id, t1.task_alloc,
            t1.task_dealloc, t2.plan_handle,
        (SELECT SUBSTRING (text, t2.statement_start_offset/2 + 1,
            (CASE WHEN statement_end_offset = -1
                THEN LEN(CONVERT(nvarchar(MAX),text)) * 2
                ELSE statement_end_offset
            END - t2.statement_start_offset)/2)
        FROM sys.dm_exec_sql_text(sql_handle)) AS query_text
FROM (SELECT session_id, request_id,
        SUM(internal_objects_alloc_page_count +
            user_objects_alloc_page_count) AS task_alloc,
        SUM(internal_objects_dealloc_page_count +
            user_objects_dealloc_page_count) AS task_dealloc
    FROM sys.dm_db_task_space_usage
    GROUP BY session_id, request_id) AS t1,
    sys.dm_exec_requests AS t2
WHERE t1.session_id = t2.session_id AND
    (t1.request_id = t2.request_id) AND t1.session_id > 50
ORDER BY t1.task_alloc DESC;
```

（2）查看执行计划(by plan-handle)

```
SELECT * FROM sys.dm_exec_query_plan(<plan-handle>)
```

3. 检查 tempdb 空闲空间的方法

```
SELECT
    SUM(user_object_reserved_page_count) * 8.192 AS UserObjectsKB,
    SUM(internal_object_reserved_page_count) * 8.192 AS InternalObjectsKB,
    SUM(version_store_reserved_page_count) * 8.192 AS VersionStoreKB,
    SUM(unallocated_extent_page_count) * 8.192 AS FreeSpaceKB
FROM sys.dm_db_file_space_usage;
```

6.4.5 阻塞诊断

1. 阻塞的检测

（1）查看 SQL Server 启用线程或工作进程遇到的累积等待时间统计值

```
SELECT TOP 10 wait_type, waiting_tasks_count AS tasks,
        wait_time_ms, max_wait_time_ms AS max_wait,
```

```
            signal_wait_time_ms AS signal
FROM sys.dm_os_wait_stats
ORDER BY wait_time_ms DESC;
```

应用程序中遇到的前 10 位等待如图 6-16 所示。

	wait_type	tasks	wait_time_ms	max_wait	signal
1	LAZYWRITER_SLEEP	7606	7607671	2046	1781
2	SQLTRACE_BUFFER_FLUSH	1901	7601625	4578	765
3	PAGEIOLATCH_SH	709	52140	921	109
4	IO_COMPLETION	422	30421	562	0
5	LCK_M_S	2	10812	8812	0
6	WRITELOG	49	4265	640	31
7	CHKPT	1	4093	4093	15
8	SLEEP_SYSTEMTASK	1	4078	4078	0
9	ASYNC_IO_COMPLETION	1	3296	3296	0
10	PAGEIOLATCH_EX	29	2171	468	0

图 6-16　应用程序中遇到的前 10 位等待

- LCK_M_S：当一个任务正等待获取一个共享锁时，LCK_M_S 等待就出现了。可以看到一个任务等待获取共享锁的最长时间是 8812ms，总共有 2 个任务在等待这个共享锁。
- LAZYWRITER_SLEEP：指的是延迟写入器线程的等待。延迟写入器线程周期性地被唤醒，并将脏页写入磁盘。因此，延迟写入器线程遇到等待是很正常的。
- PAGEIOLATCH_SH：PAGEIOLATCH_SH 等待发生在当一个任务正在等待 I/O 请求中缓冲区的一个锁存器时。锁存器请求是处于共享模式的，这种类型的长时间等待说明磁盘子系统出现了故障。
- PAGEIOLATCH_EX：当一个任务等待 I/O 请求中非缓冲区的一个锁存器时，PAGEIOLATCH_EX 等待就出现了。这个锁存器请求是处于独占模式的。
- 信号等待：是一个工作进程被授权访问资源和被列入 CPU 计划表之间的时间。一个长时间的信号等待可能意味着出现了严重的 CPU 竞争。

（2）使用系统监视计数器检测阻塞

- SQL Server:Locks:Average Wait Time（ms）指的是每个导致等待的锁请求的平均等待时间（以毫秒计）。
- SQL Server:Locks:Lock Requests/Sec 指的是锁管理器请求的新锁和锁转换的数量。
- SQL Server:Locks:Lock Wait Time（ms）指的是上一秒内等待锁的总体时间（以毫秒计）。
- SQL Server:Locks:Lock Waits/Sec 指的是不能被立即满足并要求调用者在授权予锁权限时每秒钟等待的数量。
- SQL Server:Locks:Number of Deadlocks/Sec 指的是导致一个死锁的锁请求的数量。
- SQL Server:General Statistics:Processes Blocked 指的是当前被阻塞进程的数量。
- SQL Server:Access Methods:Table Lock Escalations/Sec 指的是被升级到表级粒度的时间锁的数量。

2. 隔离和排查阻塞故障

(1) 查看哪些事务占用了锁，哪些事务被阻塞了

下面的 DMV 查询找出在任一给定时刻所有授权给当前执行事务或当前执行事务等待的锁，类似于 sp_lock。

```
SELECT request_session_id AS spid, resource_type AS rt,
    resource_database_id AS rdb,
    (CASE resource_type
        WHEN 'OBJECT'
            THEN object_name(resource_associated_entity_id)
        WHEN 'DATABASE'
            THEN ' '
        ELSE (SELECT object_name(object_id)
            FROM sys.partitions
            WHERE hobt_id=resource_associated_entity_id)
    END) AS objname,
resource_description AS rd, request_mode AS rm, request_status AS rs
FROM sys.dm_tran_locks;
```

(2) 查看阻塞的生存期和被阻塞事件执行的 SQL 信息

```
SELECT t1.resource_type,
        'database' = DB_NAME(resource_database_id),
        'blk object' = t1.resource_associated_entity_id,
        t1.request_mode, t1.request_session_id,
        t2.blocking_session_id,
        t2.wait_duration_ms,
        (SELECT SUBSTRING(text, t3.statement_start_offset/2 + 1,
            (CASE WHEN t3.statement_end_offset = -1
                THEN LEN(CONVERT(nvarchar(max),text)) * 2
                ELSE t3.statement_end_offset
            END - t3.statement_start_offset)/2)
        FROM sys.dm_exec_sql_text(sql_handle)) AS query_text,
    t2.resource_description
FROM
    sys.dm_tran_locks AS t1,
    sys.dm_os_waiting_tasks AS t2,
    sys.dm_exec_requests AS t3
WHERE
    t1.lock_owner_address = t2.resource_address AND
    t1.request_request_id = t3.request_id AND
    t2.session_id = t3.session_id;
```

(3) 显示所有索引操作统计信息

字段说明如表 6-5 所示。

表 6-5　字段说明

字　　段	字段表示意义
SELECT index_id, range_scan_count,	当访问该索引时，行和页锁便开始计数。如果数值很大，则说明这个索引被大量使用，并很可能存在竞争
row_lock_count, page_lock_count	一个请求需要等待获取一个行或一个页锁时的等待次数。同样，如果数值很大，则说明在访问该索引时存在一个竞争
FROM sys.dm_db_index_operational_stats(DB_ID('<db-name>'),	SQL Server 将该索引上的锁升级到表级的次数
OBJECT_ID('employee'), NULL, NULL);	该索引页节点上的插入、删除和更新次数

（4）查看请求锁类型

```
select   convert (smallint, req_spid) As spid,
    db_name(rsc_dbid) As '数据库名',
    object_name(rsc_objid) As '锁定对象名',
    rsc_indid As IndId,
    substring (v.name, 1, 4) As Type,
    substring (rsc_text, 1, 32) as Resource,
    substring (u.name, 1, 8) As Mode,
    substring (x.name, 1, 5) As Status

from    master.dbo.syslockinfo,
    master.dbo.spt_values v,
    master.dbo.spt_values x,
    master.dbo.spt_values u

where   master.dbo.syslockinfo.rsc_type = v.number
    and v.type = 'LR'
    and master.dbo.syslockinfo.req_status = x.number
    and x.type = 'LS'
    and master.dbo.syslockinfo.req_mode + 1 = u.number
    and u.type = 'L'
order by spid
```

6.4.6　死锁诊断

1．使用 sys.dm_tran_locks DMV 在给定的时间点探测表锁(X)

```
SELECT
    request_session_id,
    resource_type,
    DB_NAME(resource_database_id) AS DatabaseName,
    OBJECT_NAME(resource_associated_entity_id) AS TableName,
    request_mode,
    request_type,
    request_status
FROM sys.dm_tran_locks AS TL
```

```sql
        JOIN sys.all_objects AS AO
            ON TL.resource_associated_entity_id = AO.object_id
    WHERE request_type = 'LOCK'
        AND request_status = 'GRANT'
        AND request_mode IN ('X','S')
        AND AO.type = 'U'
        AND resource_type = 'OBJECT'
        AND TL.resource_database_id = DB_ID();
```

2. 防止锁升级——设置表在 1h 内防止锁升级

```sql
BEGIN TRAN
SELECT *
FROM Sales.SalesOrderDetail WITH (UPDLOCK, HOLDLOCK)
WHERE 1=0;
WAITFOR DELAY '1:00:00';
COMMIT TRAN
```

3. 显示发生 5s 以上的等待

```sql
SELECT
    WT.session_id AS waiting_session_id,
    WT.waiting_task_address,
    WT.wait_duration_ms,
    WT.wait_type,
    WT.blocking_session_id,
    WT.resource_description
FROM sys.dm_os_waiting_tasks AS WT
WHERE WT.wait_duration_ms > 5000;
```

列字段说明如表 6-6 所示。

表 6-6 列字段说明

列	备 注
Waiting_task_address	等待任务的内存地址，用于在 session 区分多任务
Session_id	与 session 的 spid 一样，用于和 sys.dm_exec_request 连接
Exec_context_id	等待任务的执行上下文 id，0 是主线程或父线程的 id
Wait_duration_ms	等待时间，以毫秒为单位
Wait_type	当前等待任务的等待类型
Resource_address	任务等待资源的内存地址，用于在 lock_owner_address 上和 sys.dm_tran_locks 连接
Blocking_task_address	阻塞任务的内存地址
Blocking_session_id	阻塞者的 session id。数-2,-3,-4 有特殊含义，详见下文
Blocking_exec_context_id	阻塞任务的执行上下文 id
Resource_description	任务等待资源的文字描述

4. 显示处于 WAIT 状态的锁

```
SELECT
    TL.resource_type,
    DB_NAME(TL.resource_database_id) as DatabaseName,
    TL.resource_associated_entity_id,
    TL.request_session_id,
    TL.request_mode,
    TL.request_status
FROM sys.dm_tran_locks AS TL
WHERE TL.request_status = 'WAIT'
ORDER BY DatabaseName, TL.request_session_id ASC;
```

5. 显示每个等待资源已授权和等待中的锁

```
SELECT
    TL1.resource_type,
    DB_NAME(TL1.resource_database_id) AS DatabaseName,
    TL1.resource_associated_entity_id,
    TL1.request_session_id,
    TL1.request_mode,
    TL1.request_status
FROM sys.dm_tran_locks as TL1
    JOIN sys.dm_tran_locks as TL2
        ON TL1.resource_associated_entity_id = TL2.resource_associated_entity_id
        AND TL1.request_status <> TL2.request_status
        AND (TL1.resource_description = TL2.resource_description
        OR (TL1.resource_description IS NULL
            AND TL2.resource_description IS NULL))
ORDER BY TL1.request_status ASC;
```

6. 返回 resource_associated_entity_id 表示的实际对象

返回 resource_associated_entity_id 表示的实际对象，不管它是 RID、Key、页还是表。

```
SELECT
    TL1.resource_type,
    DB_NAME(TL1.resource_database_id) AS DatabaseName,
    CASE TL1.resource_type
        WHEN 'OBJECT'
            THEN                                OBJECT_NAME(TL1.resource_associated_entity_id,
TL1.resource_database_id)
        WHEN 'DATABASE' THEN 'DATABASE'
        ELSE
            CASE
                WHEN TL1.resource_database_id = DB_ID() THEN
                    (SELECT OBJECT_NAME
                        (object_id, TL1.resource_database_id)
                    FROM sys.partitions
```

```
            WHERE hobt_id = TL1.resource_associated_entity_id)
        ELSE NULL
    END
END AS ObjectName,
TL1.resource_description,
TL1.request_session_id,
TL1.request_mode,
TL1.request_status
FROM sys.dm_tran_locks AS TL1
    JOIN sys.dm_tran_locks AS TL2
        ON TL1.resource_associated_entity_id =
            TL2.resource_associated_entity_id
WHERE TL1.request_status <> TL2.request_status
    AND (TL1.resource_description = TL2.resource_description
     OR (TL1.resource_description IS NULL
        AND TL2.resource_description IS NULL))
ORDER BY TL1.resource_database_id,
        TL1.resource_associated_entity_id,
        TL1.request_status ASC;
```

7．提取等待查询文本

```
SELECT
    WT.session_id AS waiting_session_id,
    DB_NAME(TL.resource_database_id) AS DatabaseName,
    WT.wait_duration_ms,
    WT.waiting_task_address,
    TL.request_mode,
    (SELECT SUBSTRING(ST.text, (ER.statement_start_offset/2) + 1,
        ((CASE ER.statement_end_offset
            WHEN -1 THEN DATALENGTH(ST.text)
            ELSE ER.statement_end_offset
         END - ER.statement_start_offset)/2) + 1)
     FROM sys.dm_exec_requests AS ER
        CROSS APPLY sys.dm_exec_sql_text(ER.sql_handle) AS ST
     WHERE ER.session_id = TL.request_session_id)
            AS waiting_query_text,
    TL.resource_type,
    TL.resource_associated_entity_id,
    WT.wait_type,
    WT.blocking_session_id,
    WT.resource_description AS blocking_resource_description,
    CASE
        WHEN WT.blocking_session_id > 0 THEN
            (SELECT ST2.text
             FROM sys.sysprocesses AS SP
                CROSS APPLY sys.dm_exec_sql_text(SP.sql_handle) AS ST2
```

```
          WHERE SP.spid = WT.blocking_session_id)
        ELSE NULL
    END AS blocking_query_text
FROM sys.dm_os_waiting_tasks AS WT
    JOIN sys.dm_tran_locks AS TL
        ON WT.resource_address = TL.lock_owner_address
WHERE WT.wait_duration_ms > 5000
    AND WT.session_id > 50;
```

解释1：
waiting_session_id 正在等待 resource_associated_entity_id 资源；

解释2：
blocking_session_id 已经获取 resource_associated_entity_id 资源；

```
    select object_name(object_id),*
    from sys.partitions
      where hobt_id = 72057594038386688
select object_name(object_id)
```

8．监视全快照隔离级别事务

```
SELECT
    transaction_id,
    session_id,
    transaction_sequence_num,
    is_snapshot,
    max_version_chain_traversed,
    elapsed_time_seconds
FROM sys.dm_tran_active_snapshot_database_transactions
ORDER BY elapsed_time_seconds DESC;
```

session_id 是指正在使用全快照隔离级别事务的 session；elapsed_time_seconds 是指事务已经被 open 的时间。

9．命令

（1）设置超时时间

设定锁请求超时，默认情况下，数据库没有超时期限（timeout_period 值为-1）
SET LOCK_TIMEOUT 1800   //ms

SELECT @@LOCK_TIMEOUT

（2）基于行版本的快照隔离模式（已提交快照和快照隔离模式）

ALTER DATABASE   替换的数据库名 SET READ_COMMITTED_SNAPSHOT ON;

ALTER DATABASE   替换的数据库名 SET ALLOW_SNAPSHOT_ISOLATION ON;
[SET TRANSACTION ISOLATION LEVEL SNAPSHOT]

（3）DMVs

```
-- dm_os_waiting_tasks.resource_address
    = dm_tran_locks.lock_owner_address
  select * from sys.dm_tran_locks
  select * from sys.dm_os_waiting_tasks


  --sql
  select * from sys.dm_exec_sql_text(sql_handle)   --sql text
  select * from sys.dm_exec_requests    --waiting
  select * from sys.sysprocesses         --blocking


  --Object Name
  SELECT OBJECT_NAME(object_id),*
  FROM sys.partitions
  WHERE hobt_id = dm_tran_locks.resource_associated_entity_id

  OBJECT_NAME(dm_tran_locks.resource_associated_entity_id)
```

10. 跟踪标记 1204

```
-1 表示不仅针对当前连接，而且针对以后新建立的连接
DBCC TRACEON(1204,3605,-1)
GO


DBCC TRACEOFF(1204,3605)
GO

DBCC    TRACEOFF(1204,-1)
```

11. 查看锁信息（存储过程 1）

```
Create Proc sp_us_lockinfo
AS
BEGIN
    SELECT
        DB_NAME(t1.resource_database_id) AS [数据库名],
        t1.resource_type AS [资源类型],
  --    t1.request_type AS [请求类型],
        t1.request_status AS [请求状态],
  --    t1.resource_description AS [资源说明],
        CASE t1.request_owner_type WHEN 'TRANSACTION' THEN '事务所有'
                                   WHEN 'CURSOR' THEN '游标所有'
                                   WHEN 'SESSION' THEN '用户会话所有'
                                   WHEN 'SHARED_TRANSACTION_WORKSPACE' THEN '事务工作区的共享所有'
```

```
                              WHEN 'EXCLUSIVE_TRANSACTION_WORKSPACE' THEN'
事务工作区的独占所有'
                                          ELSE ''
            END AS [拥有请求的实体类型],
            CASE WHEN T1.resource_type = 'OBJECT'
                THEN OBJECT_NAME(T1.resource_ASsociated_entity_id)
                ELSE   T1.resource_type+':'+ISNULL(LTRIM(T1.resource_ASsociated_entity_id),'')
                END AS [锁定的对象],
            t4.[name] AS [索引],
            t1.request_mode AS [锁定类型],
            t1.request_session_id AS [当前 spid],
            t2.blocking_session_id AS [锁定 spid],
            --t3.snapshot_isolation_state AS [快照隔离状态],
            t3.snapshot_isolation_state_desc AS [快照隔离状态描述],
            t3.is_read_committed_snapshot_on AS [已提交读快照隔离]
        FROM
            sys.dm_tran_locks AS t1
        left join
            sys.dm_os_waiting_tasks AS t2
        ON
            t1.lock_owner_address = t2.resource_address
        left join
            sys.databases AS t3
        ON t1.resource_database_id = t3.database_id
        left join
            (
              SELECT rsc_text,rsc_indid,rsc_objid,b.[name]
            FROM
                sys.syslockinfo a
            JOIN
                sys.indexes b
            ON a.rsc_indid = b.index_id and b.object_id = a.rsc_objid) t4
            ON t1.resource_description = t4.rsc_text
    END
    GO
```

调用 exec sp_us_lockinfo

12. 查看锁信息（存储过程 2）

```
use master
go
create procedure sp_who_lock
as
begin
declare @spid int,@bl int,
 @intTransactionCountOnEntry       int,
 @intRowcount                      int,
 @intCountProperties               int,
```

```
@intCounter                int
create table #tmp_lock_who (
id int identity(1,1),
spid smallint,
bl smallint)
IF @@ERROR<>0 RETURN @@ERROR
insert into #tmp_lock_who(spid,bl) select   0 ,blocked
from (select * from sysprocesses where    blocked>0 ) a
where not exists(select * from (select * from sysprocesses
where    blocked>0 ) b
where a.blocked=spid)
union select spid,blocked from sysprocesses where    blocked>0
IF @@ERROR<>0 RETURN @@ERROR
-- 找到临时表的记录数
select       @intCountProperties = Count(*),@intCounter = 1
from #tmp_lock_who
IF @@ERROR<>0 RETURN @@ERROR
if     @intCountProperties=0
select '现在没有阻塞信息' as message
-- 循环开始
while @intCounter <= @intCountProperties
begin
-- 取第一条记录
select       @spid = spid,@bl = bl
from #tmp_lock_who where Id = @intCounter
begin
if @spid =0
select '引起数据库阻塞的是: '+ CAST(@bl AS VARCHAR(10))
+ '进程号,其执行的 SQL 语法如下'
else
select '进程号 SPID：'+ CAST(@spid AS VARCHAR(10))+ '被'
+ '进程号 SPID：'+ CAST(@bl AS VARCHAR(10)) +'阻塞,其当前进程执行的 SQL 语法如下'
DBCC INPUTBUFFER (@bl )
end
-- 循环指针下移
set @intCounter = @intCounter + 1
end
drop table #tmp_lock_who
return 0
end

sp_who_lock
```

6.4.7 排除故障

1．207 错误提示

执行 select * from sysmessages where error=207

2．查看连接数

(1) SP_WHO 'sa'

(2) select count(*) as ConnNum from master..sysprocesses where db_name(dbid)='Performance'

(3) select * from sysprocesses where dbid in (select dbid from sysdatabases where name='Performance')

(4) select @@max_connections

6.4.8 信息查询

1．查看数据库是否启用 AWE

```
EXEC sp_configure 'show advanced options', 1
RECONFIGURE
GO

EXEC sp_configure 'awe enabled'
GO
```

2．查看 SQL Server 通过 AWE 机制分配了多少内存

```
select
    sum(awe_allocated_kb) / 1024 as [AWE allocated, Mb]
from
sys.dm_os_memory_clerks
```

6.4.9 存储引擎

1．查看数据库信息（实例/数据库/文件信息）

数据库实例的概要情况：

```
SELECT * FROM SYS.SERVERS
WHERE SERVER_ID = 0
-- 兼容性视图 SELECT * FROM SYS.SYSSERVERS
```

各个数据库的详细信息：

```
SELECT * FROM SYS.DATABASES
-- 兼容性视图 SELECT * FROM SYS.SYSDATABASES
```

文件组的详细信息：

```
SELECT * FROM SYS.FILEGROUPS
-- 兼容性视图 SELECT * FROM SYS.SYSFILEGROUPS
```

各个数据库文件的详细信息：

```
SELECT * FROM SYS.MASTER_FILES
-- 兼容性视图 SELECT * FROM SYS.SYSALTFILES
```

当前数据库文件的详细信息：

```
SELECT * FROM SYS.DATABASE_FILES
    -- 兼容性视图 SELECT * FROM SYS.SYSFILES
```

数据空间的详细情况，可以是文件组或分区方案：

```
SELECT * FROM SYS.DATA_SPACES
```

2．查看数据库信息（表/对象）

事实上几乎所有的用户对象都出自 SYS.OBJECTS 表。

```
SELECT * FROM SYS.OBJECTS
WHERE type_desc = 'USER_TABLE' AND NAME='TEST'
    -- 兼容性视图 SYSOBJECTS
```

如果要查询与该表相关的其他所有对象，则可以执行以下语句：

```
SELECT * FROM SYS.OBJECTS
WHERE type_desc = 'USER_TABLE' AND NAME='TEST' OR
       parent_object_id in
        ( SELECT object_id FROM SYS.OBJECTS
            WHERE type_desc = 'USER_TABLE' AND NAME='TEST')
```

表字段详细信息，可以查询出相关 column_id。

```
SELECT * FROM SYS.COLUMNS
WHERE OBJECT_ID = 309576141
    -- 兼容性视图 SYSCOLUMNS
```

表索引详细情况，可以清楚地看到存在两个索引。

```
SELECT * FROM SYS.INDEXES WHERE OBJECT_ID = 309576141
    -- 兼容性视图 SYSINDEXES
```

表分区情况，数据库中所有表和索引的每个分区在表中各对应一行。此处可以看到该表有两个分区，聚集索引即表本身，还有一个是 name 的非聚集索引，partition_id 即分区的 ID。hobt_id 包含此分区的行的数据堆或 B 树的 ID。

```
SELECT * FROM SYS.PARTITIONS WHERE OBJECT_ID = 309576141
```

分配单元情况，数据库中的每个分配单元都在表中占一行。该表只有和 SYS.PARTITIONS 配合使用才有意义。

```
SELECT * FROM SYS.ALLOCATION_UNITS
```

SYS.ALLOCATION_UNITS 和 SYS.PARTITIONS 一起使用，能够反映出某个对象的页面分配和使用情况。

```
SELECT * FROM SYS.ALLOCATION_UNITS U,SYS.PARTITIONS P
WHERE U.TYPE IN ( 1 , 3 ) AND U.CONTAINER_ID = P.HOBT_ID AND P.OBJECT_ID = 309576141
UNION ALL
```

```
SELECT * FROM SYS.ALLOCATION_UNITS U,SYS.PARTITIONS P
WHERE U.TYPE = 2 AND U.CONTAINER_ID = P.PARTITION_ID AND P.OBJECT_ID = 309576141
```

返回每个分区的页和行计数信息：

```
SELECT * FROM SYS.DM_DB_PARTITION_STATS WHERE OBJECT_ID = 309576141
```

返回索引的详细字段情况：

```
SELECT * FROM SYS.INDEX_COLUMNS WHERE OBJECT_ID = 309576141
    -- 兼容性视图 SYSINDEXKEYS
```

以下为根据某个索引名称获取其相关字段的语句：

```
DECLARE @index_field_names VARCHAR( 500 )
SET @index_field_names = '';
SELECT @index_field_names = @index_field_names + c.name + ','
    FROM SYS.INDEX_COLUMNS a,SYS.INDEXES b,SYS.COLUMNS c
WHERE a.object_id = b.object_id AND a.index_id = b.index_id
    AND a.object_id = c.object_id AND a.column_id = c.column_id
    AND b.name = 'IX_test'
ORDER BY a.index_column_id
SET @index_field_names = LEFT (@index_field_names, LEN (@index_field_names) - 1 )
PRINT @index_field_names
```

CHECK 约束，数据来源 sys.objects.type = 'C'。

```
SELECT * FROM SYS.CHECK_CONSTRAINTS WHERE OBJECT_ID = 309576141
    -- 兼容性视图 SYSCONSTRAINTS
```

数据来源 sys.objects.type = D。

```
SELECT * FROM SYS.DEFAULT_CONSTRAINTS WHERE OBJECT_ID = 309576141
    -- 兼容性视图 SYSCONSTRAINTS
```

主键或唯一约束,数据来源 sys.objects.type PK 和 UQ。

```
SELECT * FROM SYS.KEY_CONSTRAINTS WHERE OBJECT_ID = 309576141
    -- 兼容性视图 SYSCONSTRAINTS
```

外键，数据来源 sys.object.type = F。

```
SELECT * FROM SYS.FOREIGN_KEYS WHERE OBJECT_ID = 309576141
    -- 兼容性视图 SYSREFERENCES
```

触发器：

```
SELECT * FROM SYS.TRIGGERS WHERE OBJECT_ID = 309576141
```

注释：

```
SELECT * FROM SYS.SQL_MODULES
```

-- 兼容性视图 SYSCOMMENTS

数据库用户表：

   SELECT * FROM SYS.DATABASE_PRINCIPALS
   -- 兼容性视图 SYSUSERS

数据库数据类型表：

   SELECT * FROM SYS.TYPES
   -- 兼容性视图 SYSTYPES

3. 查看数据库信息（对象存储视图）

根据表名称查询出 object_id。

   SELECT name,object_id,parent_object_id,type_desc
      FROM SYS.OBJECTS WHERE NAME = 'TEST'   --object_id: 389576426

再查询相关索引视图，可以清楚地看到索引视图中包含两条索引记录，即聚集索引和非聚集索引。

   SELECT object_id,name,index_id,type,type_desc
      FROM SYS.INDEXES WHERE OBJECT_ID = 389576426

再查询相关分区视图，可以看到分区视图中包含两条记录，即聚集索引和非聚集索引。

   SELECT partition_id,object_id,index_id,partition_number,hobt_id,rows
      FROM SYS.PARTITIONS WHERE OBJECT_ID = 389576426

再查询分配单元视图，可以看到分配单元视图中包含三条记录，即聚集索引和非聚集索引以及 LOB 数据。

   SELECT allocation_unit_id,type,type_desc,container_id,
         data_space_id,total_pages,used_pages,data_pages
      FROM
      (
         SELECT * FROM SYS.ALLOCATION_UNITS U,SYS.PARTITIONS P
         WHERE U.TYPE IN (1 , 3) AND U.CONTAINER_ID = P.HOBT_ID
            AND P.OBJECT_ID = 389576426
         UNION ALL
         SELECT * FROM SYS.ALLOCATION_UNITS U,SYS.PARTITIONS P
         WHERE U.TYPE = 2 AND U.CONTAINER_ID = P.PARTITION_ID
            AND P.OBJECT_ID = 389576426
) A

最后再查询 system_internals_allocation_units 视图，可以看到该视图与分配单元视图基本类似，但多了三个页面地址：

   SELECT allocation_unit_id,type,type_desc,container_id,filegroup_id,
         total_pages,used_pages,data_pages,

```
            Performance.dbo.f_get_page(first_page) first_page_address,
            Performance.dbo.f_get_page(root_page) root_address,
                    Performance.dbo.f_get_page(first_iam_page) IAM_address
        FROM
        (
            SELECT * FROM sys.system_internals_allocation_units U,SYS.PARTITIONS P
                WHERE U.TYPE IN ( 1 , 3 ) AND U.CONTAINER_ID = P.HOBT_ID
                    AND P.OBJECT_ID = 389576426
            UNION ALL
            SELECT * FROM sys.system_internals_allocation_units U,SYS.PARTITIONS P
                WHERE U.TYPE = 2 AND U.CONTAINER_ID = P.PARTITION_ID
                    AND P.OBJECT_ID = 389576426
        ) A


[
    CREATE FUNCTION [dbo].f_get_page(@page_num BINARY( 6 ))
        RETURNS VARCHAR( 11 )
        AS
        BEGIN
            RETURN(CONVERT(VARCHAR( 2 ),(CONVERT( INT ,SUBSTRING(@page_num, 6 , 1 )) * POWER( 2 , 8 )) +
                (CONVERT( INT ,SUBSTRING(@page_num, 5 , 1 )))) + ' :'+
                CONVERT(VARCHAR( 11 ),
                (CONVERT( INT ,SUBSTRING(@page_num, 4 , 1 )) * POWER( 2 , 24 )) +
                (CONVERT( INT ,SUBSTRING(@page_num, 3 , 1 )) * POWER( 2 , 16 )) +
                (CONVERT( INT ,SUBSTRING(@page_num, 2 , 1 )) * POWER( 2 , 8 )) +
                (CONVERT( INT ,SUBSTRING(@page_num, 1 , 1 )))))
        END
]
```

4．查看页面结构

（1）SQL Internals Viewer 工具

SQL Internals Viewer 工具中的页面类型及其描述如下。

① Data page：堆表和聚集索引的叶子节点数据。

② Index page：聚集索引的非叶子节点和非聚集索引的所有索引记录。

③ Sort page：排序时所用到的临时页，排序中间操作存储数据用的。

④ GAM page：全局分配映射（Global Allocation Map，GAM）页面，这些页面记录了哪些区已经被分配并用作何种用途。

⑤ SGAM page：共享全局分配映射（Shared Global Allocation Map，GAM）页面，这些页面记录了哪些区当前被用作混合类型的区，并且这些区需含有至少一个未使用的页面。

⑥ IAM page：有关每个分配单元中表或索引所使用的区的信息。

⑦ PFS page：有关页分配和页可用空间的信息。

⑧ boot page：记录了关于数据库的信息，仅存于每个数据库的第 9 页。

⑨ file header page：记录了关于数据库文件的信息，存于每个数据库文件的第 0 页。
⑩ DCM page：记录自从上次完全备份以来的数据改变的页面，以便进行差异备份。
⑪ BCM page：有关每个分配单元中自最后一条 BACKUP LOG 语句之后的大容量操作所修改的区的信息。

（2）DBCC Page

DBCC Page ({dbid|dbname},filenum,pagenum[,printopt])

具体参数描述如下：
dbid：包含页面的数据库 ID。
dbname：包含页面的数据库名称。
filenum：包含页面的文件编号。
pagenum：文件内的页面。
printopt：可选的输出选项，选用以下值中的一个：
    0：默认值，输出缓冲区的标题和页面标题。
    1：输出缓冲区的标题、页面标题(分别输出每一行)，以及行偏移量表。
    2：输出缓冲区的标题、页面标题(整体输出页面)，以及行偏移量表。
    3：输出缓冲区的标题、页面标题(分别输出每一行)，以及行偏移量表。每一行后分别列出它的列值。
如果想看到这些输出的结果，还需要设置 DBCC TRACEON(3604)。

```
DBCC TRACEON( 3604 )
DBCC PAGE(Performance, 1 , 2 , 1 )    //查看 GAM 页信息。
DBCC PAGE(Performance, 1 , 3 , 1 )    //查看 SGAM 页信息。
DBCC PAGE(Performance, 1 , 2 , 2 )    //查看 GAM 页信息和整体输出页面。
DBCC PAGE(Performance, 1 , 3 , 2 )    //查看 SGAM 页信息和整体输出页面。
DBCC PAGE(Performance, 1 , 2 , 3 )    //查看 GAM 页信息及相应列值。
DBCC PAGE(Performance, 1 , 3 , 3 )    //查看 SGAM 页信息及相应列值。
DBCC PAGE(Performance, 1 , 2 , 1 ) WITH TABLERESULTS   //以表格形式查看 SGAM 页信息及相应列值。
DBCC PAGE(Performance, 1 , 3 , 1 ) WITH TABLERESULTS   //以表格形式查看 SGAM 页信息及相应列值。
```

例如：

```
DBCC TRACEON( 3604 )
DBCC PAGE(10, 1 , 1794 , 1 )
select object_name(149575571)
```

5．PFS 页面结构

```
dbcc page(Performance,1,1,2)
```

第 0 个 bit 为保留字节，始终为 0。
第 1 个 bit 表示该页面是否已分配。GAM 页用来管理区是否已分配，但一个区包含 8 个页面，所以用该 bit 来准确定位该区的某个页面是否已分配出去。

第 2 个 bit 表示该页面是否为混合分区的一个页面。

第 3 个 bit 表示该页面是否为一个 IAM 页面。

第 4 个 bit 表示该页面中是否包含幻影或已删除记录，这有助于 SQL Server 定期清理幻影或已删除记录。

第 5~7 个 bit 表示该页面的空间使用率情况：
- 0：表示该页面为空。
- 1：表示该页面已使用 1%~50%。
- 2：表示该页面已使用 51%~80%。
- 3：表示该页面已使用 81%~95%。
- 4：表示该页面已使用 96%~100%。

6．IAM 页面结构

```
    SELECT total_pages,used_pages,data_pages,
           first_page,root_page,first_iam_page,
           Performance.dbo.f_get_page(first_page) first_page_address,
           Performance.dbo.f_get_page(root_page) root_address,
           Performance.dbo.f_get_page(first_iam_page) IAM_address
    FROM sys.system_internals_allocation_units
WHERE container_id IN ( SELECT partition_id FROM sys.partitions
                        WHERE object_id in ( SELECT object_id   FROM sys.objects
                                             WHERE name IN ( 'Workload')))


    --DBCC TRACEON( 3604 )
dbcc page(Performance, 1 , 93 , 3 )    --93：  IAM_address
```



```
    [
           CREATE FUNCTION [dbo].f_get_page(@page_num BINARY( 6 ))
              RETURNS VARCHAR( 11 )
              AS
              BEGIN
                   RETURN(CONVERT(VARCHAR( 2 ),(CONVERT( INT ,SUBSTRING(@page_num, 6 ,
1 )) * POWER( 2 , 8 )) +
                              (CONVERT( INT ,SUBSTRING(@page_num, 5 , 1 )))) + ' :'+
                         CONVERT(VARCHAR( 11 ),
                           (CONVERT( INT ,SUBSTRING(@page_num, 4 , 1 )) * POWER( 2 , 24 )) +
                           (CONVERT( INT ,SUBSTRING(@page_num, 3 , 1 )) * POWER( 2 , 16 )) +
                           (CONVERT( INT ,SUBSTRING(@page_num, 2 , 1 )) * POWER( 2 , 8 )) +
                           (CONVERT( INT ,SUBSTRING(@page_num, 1 , 1 )))))
          END
    ]
```

7. DBCC IND

TRUNCATE TABLE tablepage;
INSERT INTO tablepage EXEC (' DBCC IND(Performance,tablename,1)');
SELECT
  PagePID,IAMPID,ObjectID,IndexID,Pagetype,IndexLevel,
  NextPagePID,PrevPagePID
FROM tablepage

[
    DBCC IND({'dbname'|dbid},{'objectname'|objectID},
  {nonclustered indid|1|0|-1|-2}[,partition_number])
  {'dbname'|dbid}表示数据库名或者数据库 ID
  {'objectname'|objectID}表示对象名或者对象 ID
  {nonclustered indid|1|0|-1|-2}表示显示行内数据分页及指定对象的行内 IAM 分页信息
  1：显示所有分页的信息，包括 IAM 分页、数据分页、所有存在的 LOB 分页、行溢出页和索引分页。
  -1：显示所有 IAM、数据分页，及指定对象上全部索引的索引分页。
  -2：显示指定对象的所有 IAM 分页。
  nonclustered indid：显示所有的 IAM、数据分页以及一个索引的索引分页信息。
  {partition_number}：可选，为了与中间的 DBCC IND 命令向前兼容，它指定了一个特定分区号，如果不指定，则显示所有分区的信息。
]
[
CREATE TABLE tablepage
(
    PageFID   TINYINT,
    PagePID   INT ,
    IAMFID   TINYINT,
    IAMPID   INT ,
    ObjectID   INT ,
    IndexID   TINYINT,
    Partition_Number TINYINT,
    PartitionID   BIGINT,
    iam_chain_type VARCHAR(30),
    PageType   TINYINT,
    IndexLevel   TINYINT,
    NextPageFID   TINYINT,
    NextPagePID   INT ,
    PrevPageFID   TINYINT,
    PrevPagePID   INT
);
GO
]
[

PageFID：页面的文件号。
PagePID：页面号。
IAMFID：IAM 所在文件号，如果是 IAM 页面则此项为空。
IAMPID：对应的 IAM 页面号，对于 IAM 为 NULL。
ObjectID：页面所属对象 ID。
IndexID：索引的 ID 号。
PartitionNumber：分区号。
PartitionID：分区 ID。
iam_chain_type：有 3 种类型（2005 及以上版本），即 IN_ROW_DATA、LOB_DATA 和 ROW_OVERFLOW_DATA。
PageType（页面类型）：

    1 - data page
    2 - index page
    3 and 4 - text pages
    5 - GAM page
    6 - SGAM page
    7 - IAM page
    8 - PFS page

IndexLevel：对应页面表头部分的 m_level，1 为当前页面在索引中的级数。

8．获取表的页面信息

```
-- 获取该表相应的页面信息
SELECT A . NAME TABLE_NAME , B . NAME INDEX_NAME , B . INDEX_ID
  FROM SYS . OBJECTS A , SYS . INDEXES B
  WHERE A . OBJECT_ID = B . OBJECT_ID AND A . NAME = 'testHeapIndex'
TRUNCATE TABLE tablepage ;
INSERT INTO tablepage EXEC ( 'DBCC IND(Performance,testHeapIndex,0)' );
INSERT INTO tablepage EXEC ( 'DBCC IND(Performance,testHeapIndex,2)' );
INSERT INTO tablepage EXEC ( 'DBCC IND(Performance,testHeapIndex,3)' );
SELECT
  b . name table_name ,
  CASE WHEN c . type = 0 THEN ' 堆'
       WHEN c . type = 1 THEN ' 聚集'
       WHEN c . type = 2 THEN ' 非聚集'
       ELSE ' 其他'
  END index_type ,
  c . name index_name ,
  PagePID , IAMPID , ObjectID , IndexID , Pagetype , IndexLevel ,
  NextPagePID , PrevPagePID
  FROM tablepage a , sys . objects b , sys . indexes c
  WHERE A . ObjectID = b . object_id
    AND A . ObjectID = c . object_id
    AND a . IndexID = c . index_id
```

```sql
-- 获取该表的 root 页面地址，聚集索引
SELECT c.name, a.type_desc, d.name,
       total_pages, used_pages, data_pages,
       Performance.dbo.f_get_page(first_page) first_page_address,
       Performance.dbo.f_get_page(root_page) root_address,
       Performance.dbo.f_get_page(first_iam_page) IAM_address
FROM sys.system_internals_allocation_units a, sys.partitions b, sys.objects c, sys.indexes d
WHERE a.container_id = b.partition_id and b.object_id = c.object_id
  AND d.object_id = b.object_id    AND d.index_id = b.index_id
  AND c.name in ('表名')
```

9. 转换页面地址为页号

```sql
CREATE PROC sp_FPSInfo
@FORWARDING_STUB BINARY(8)
AS
SELECT
   CAST(
      CONVERT(INT,SUBSTRING(@FORWARDING_STUB,6,1)) * POWER(2,8)
      + CONVERT(INT,SUBSTRING(@FORWARDING_STUB,5,1))
         AS VARCHAR)+'       :' --File_num
   +CAST(
      (CONVERT(INT,SUBSTRING(@FORWARDING_STUB,4,1)) * POWER(2,24))
      + (CONVERT(INT,SUBSTRING(@FORWARDING_STUB,3,1)) * POWER(2,16))
      + (CONVERT(INT,SUBSTRING(@FORWARDING_STUB,2,1)) * POWER(2,8 ))
      + (CONVERT(INT,SUBSTRING(@FORWARDING_STUB,1,1))) AS VARCHAR)
      +' :'  --Page_id
   +CAST(CAST(SUBSTRING(@FORWARDING_STUB,8,1) * POWER(2,8 ) +
         + SUBSTRING(@FORWARDING_STUB,7,1) AS INT ) AS VARCHAR)
      AS 'FILE_NUM:PAGE_ID:SLOT_ID'
GO

EXEC   sp_us_FPSinfo 0x0000000010000200 --16 进制地址
```

10. 查看分区情况

```sql
select * from sys.partition_functions
select * from sys.partition_range_values
select * from sys.partition_schemes
```

第 7 章
基于 Web 技术的性能优化方案

本章内容

- Web 技术优化方案
- 网络瓶颈诊断

7.1 Web 技术优化方案

本节主要介绍一下 ASP.NET 常用的一些全局优化方案。很多方案是通用的,不仅局限于这一个平台。

7.1.1 发布时要关闭调试模式

设置 compilation debug="true",将调试符号插入已编译的页面中。但这会影响性能,因此只在开发过程中将此设置为 true,发布时设置为 false。

7.1.2 服务器和客户缓存利用

缓存包括服务器端和客户端缓存,充分利用缓存可以减少服务器压缩,提高服务器整体性能。

7.1.3 启用 GZIP 压缩功能

以下是在 IIS(这里是 6.0 的修改方法)启用 GZIP 功能的一个案例,本案例将对网站中的一些静态资源和应用程序文件类型启用 GZIP 压缩。

GZIP 压缩功能配置分为以下两步:

(1) 在 IIS 的 "网站" 上单击鼠标右键,选择 "属性" 命令打开 "网站属性" 对话框,单击 "服务" 选项卡,选中 "HTTP 压缩" 中的两个复选框。

(2) 修改 c:\WINDOWS\system32\inetsrv\MetaBase.xml 文件。

修改 HcFileExtensions 的值为:

```
HcFileExtensions="htm html txt js css"
```

修改 HcScriptFileExtensions 的值为:

```
HcScriptFileExtensions="asp aspx dll exe ashx axd"
```

7.1.4 对站点中的静态资源精简与压缩

7.1.3 中启用的 GZIP 压缩是 IIS 对页面进行的压缩,它是运行时的 "动态压缩",压缩的力度还不够。另外,GZIP 对图片默认不进行压缩,因为大多数格式类型的图片都是已经经过压缩的,压缩比比较低。

这里主要介绍基于 GZIP,且在 GZIP 压缩之前的 "静态压缩",主要包括对静态资源(JavaScript 文件、CSS 文件、Image 图片)进行精简和(或)压缩。

JavaScript 文件可以使用 JSMin 工具进行精简。JSMin 可以使文件压缩 20% 以上,不同的脚本文件压缩率并不相同,这与脚本中注释的各种格式字符数量也有关系。

图片压缩可以使用 Professional Image Optimizer 工具,此工具可以支持常用的多种文件格式(jpg、gif、png、tiff),且支持批量压缩。使用图像处理软件生成的图片可以通过此工具再进行压缩,压缩后可能分辨率稍差,但从网站的使用及性能角度考虑,取个恰当的分辨率和压缩率，可以使用户得到最佳的性能体验。另外，此工具对较大的图片及较小的（小于1KB）的图片均能够继续压缩，在我的优化过程中对各种大小尺寸的图片均采用此工具压缩处理。其中有个52KB的图片经过压缩处理后仅为16KB；小于1KB的图片（940B）压缩后仅为334B，且压缩前后图片看上去效果差不多。如果对全站点所有图片都进行压缩处理的话，能够取得较大的性能提升。

CSS与JavaScript一样，在发布之前也需要进行精简，且精简并发布后，仍然可以被IIS进一步采用GZIP压缩。CSS精简/压缩工具也有很多，包括一些在线压缩程序等。

7.1.5 JavaScript/CSS 输出位置规范

针对CSS和JavaScript两种类型文件，一般优化规范如下：

1. CSS要放到页面头部，以免页面呈现过程中产生回流现象。
2. JavaScript放到页面尾部。虽然浏览器有多个并发线程，可以同时下载多个页面资源，但JavaScript比较特殊，它的下载和执行是独占时间的。举个例子，IE8有6个并发线程，如果第一个线程在下载JavaScript，则其他线程会空闲等待此文件下载并执行完成后，再下载其他资源。因此尽量把JavaScript放到页面底部可以使该脚本文件最后执行。

此外，JavaScript执行过程与后台异步非常相似，把JavaScript放到最后执行虽然页面加载的总时间相同，但对于用户来说，最后JavaScript的执行时间感觉不到，从而感觉页面打开时间会变快些。

下面是一个页面结构图：

```
<html xmlns="http://www.w3.org/1999/xhtml">
<head runat="server">
    <title></title>
</head>
<body>
    <form id="form1" runat="server">
        ①
…… //页面中 dom element etc
        …… //页面中 dom element etc
        ②
    </form>
</body>
</html>
```

在页面结构中，②的位置要比①的位置好很多。在用C#输出时要使用：

    Page.ClientScript.RegisterStartupScript();   //在②位置输出

代替语句：

    Page.ClientScript.RegisterClientScriptBlock();   //在①位置输出

在某些特殊情况下，用代码很难实现脚本在页面底部输出。这时可以使用defer关键字，让脚本延迟执行，如下：

    <script type="text/javascript" defer></script>

以上写法是直接在静态页面中写法，也可以在服务器端输出，如下：

Page.ClientScript.RegisterClientScriptBlock(this.GetType(),
"SearchControlIntelligentScript","<script type='text/javascript' src='" + strJSPath
+ "' defer></script>", false);

注意：defer 是 IE 浏览器特有的功能。

另外，对于极少数特殊的脚本，必须在页面特定位置输出的，如含有 document.write 方法的，则不能使用此方案进行优化。

7.1.6 减少页面请求

页面在加载过程中，会单独对页面中包含的每一个资源（JavaScript，CSS，图片）请求服务器获取，因此应该尽量减少页面请求数。

对于 JavaScript 和 CSS 可以通过资源合并的方式实现；对于图片，一些成组的图片可以使用 CSS Sprites 实现。这两种方式都能够减少页面中的请求数量。

CSS Sprites 的工具条按钮在实现时，不需要单独做每个图片，仅需要把所有工具图案做到 1 个图片上，然后通过 CSS 样式实现每个工具条控件对应图片的相对位置即可。下面是一个 CSS Sprites 例子，显示包含 5 个按钮的工具条样式定义如下：

```
<style type="text/css">
    .tool1  {background-image:url(tools.png);width:10px;height:10px;background-position:0   0;display:inline;float:left;}
    .tool2  {background-image:url(tools.png);width:10px;height:10px;background-position:-11 0;display:inline;float:left;}
    .tool3  {background-image:url(tools.png);width:10px;height:10px;background-position:-21 0;display:inline;float:left;}
    .tool4  {background-image:url(tools.png);width:10px;height:10px;background-position:-31 0;display:inline;float:left;}
    .tool5  {background-image:url(tools.png);width:10px;height:10px;background-position:-41 0;display:inline;float:left;}
</style>
```

工具条控件定义如下：

```
<div id="工具条">
    <a href="javascript:alert('tool1')" ><span class="tool1"></span></a>
    <a href="javascript:alert('tool2')" ><span class="tool2"></span></a>
    <a href="javascript:alert('tool3')" ><span class="tool3"></span></a>
    <a href="javascript:alert('tool4')" ><span class="tool4"></span></a>
    <a href="javascript:alert('tool5')" ><span class="tool5"></span></a>
</div>
```

工具条的每个子控件都通过 class 指定不同的样式，实际上是通过样式的 background-position 来指定图片上每个子控件的坐标偏移量来实现的。

采用这种巧妙的方式可以减少页面中的请求数量，减少页面的加载时间。

7.1.7 禁用服务器控件的视图状态

对不需要进行回发处理的控件，禁用视图状态。在《庖丁解牛：纵向切入 ASP.NET 3.5 控件和组件开发技术》一书的第 6 章和我的博客上详细讲解了视图状态：

http://blog.csdn.net/chengking/article/details/3680629

7.1.8 定制仅满足特定功能的自定义控件

自定义控件的好处是：由于是自己开发的控件，所以当出现瓶颈时，有优化的可能性。如果使用的是第三方控件，由于第三方控件一般不会发布源代码，遇到瓶颈时即使确认是第三方控件的问题，也无法修改代码进行优化。

7.1.9 优化方案提升数据

针对大多数产品，总结了一些经验数据，如表 7-1 所示。

表 7-1 性能提升数据

优 化 方 案	优 化 效 果
减少 HTTP Request 次数	做到极致，效率可以提升 60%
JavaScript 合并	充分利用浏览器缓存，效率可以提升 40%
使用 CDN	效率可提升 20%
完整缓存页面	效率可提升 75%～85%
启用 gzip 压缩	效率可提升 66%
启用 deflate	效率可提升 60%
JavaScript 压缩——JSMin	效率可提升 21%
JavaScript 压缩——Dojo Compressor	效率可提升 25%
JSMin+gzip	效率可提升 78%

注：这些仅是一些针对大多数产品的经验数据，仅供参考。

7.2 网络瓶颈诊断

7.2.1 各种网速测试方法

在客户现场，客户会说系统运行非常慢，作为性能工程师，首先要判断是网络瓶颈导致，还是其他软件或硬件出现故障。下面就介绍几种常用的网络速度检测方法。

有一次，刚刚在一个客户现场装上一套系统管理软件，用于 Internet，客户反映运行比较慢。性能工程师需要测试一下客户机连接服务器的准确速度。

最简单的方法是使用浏览器从服务器下载一个文件，看一下下载速度。但浏览器下载窗口显示的速度值经常不精确，只能了解大概。

方法一：使用网速检测工具

我最常用的是 DU Meter 或 360 的网速监视器，这些工具能够检测出当前计算机哪些进程各自占用了多少流量，且同时支持上行和下行速度检测。DU Meter 网速监视器如图 7-1 所示：

图 7-1 DU Meter 网速监视窗口

可以看到当前下行速度为 135.6KB/s（左边值），上行速度 2.9KB/s（右边值），上行速度监视值用于从本机上传文件到其他服务器时。

笔者在客户现场使用最多的是 DU Meter，因为 DU Meter 网速检测值比较准确。

方法二：使用 Perfmon 网络接口对象

选择"开始"→"运行"，输入命令"perfmon"，则会弹出"性能"窗口，然后在右侧面板空白处点击鼠标右键，选择"添加计数器"，打开"添加计数器"对话框，如图 7-2 所示。

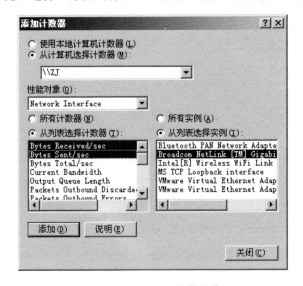

图 7-2 DU Meter 网速监视窗口

在"性能对象"下拉框中选择"Network Interface"对象，在"从列表选择计数器"列表框中选择"Bytes Received/sec"和"Bytes Sent/sec"，在"从列表选择实例"列表框中选择"Broadcom NetLink [TM] Gigabit Etherent"（可以选择当前网络接口设备），即可将这两个计数器添加到监视窗口，可以看到当前下行速度为约 139KB，如图 7-3 所示。

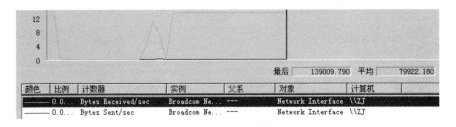

图 7-3 Perfmon——Network Interface

方法三：使用任务管理器"联网"功能

打开任务管理器，选择"联网"选项卡。在上方会显示所有网络设备网络带宽利用率图形,下方会显示各个网络接口设备的速度,如图 7-4 所示。

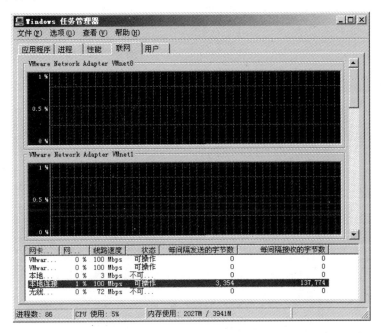

图 7-4 任务管理器窗口

每间隔发送的字节数即为当前上行速度,每间隔接收的字节数即为当前下行速度。如图 7-4 所示,当前下行速度约为 137KB/s。

默认情况下,列每间隔发送的字节数和每间隔接收的字节数是不显示的,需要从任务管理器菜单中设置。

此外,还有其他一些工具,如 360 安全卫士等也提供了网速检测功能。还有很多在线网络检测工具,这些工具的优势是可以同时测试当前计算机的上行和下行最大速度。

7.2.2 网络瓶颈诊断

在 7.2.1 中介绍了怎么检测客户端连接服务器的网络速度。通过网速测试工具可以很容易地测试出客户端机器连接服务端机器的速度。比如,测试到客户端连接服务器的实际速度为 50KB/s,网速比较慢。对于此数据只能确定以下几点:

- 可能是客户端有瓶颈。
- 可能是服务端有瓶颈。
- 也可能是客户端和服务端都有瓶颈。

但不能具体确认瓶颈到底在哪。为了能够正确定位瓶颈,可以预先在 Internet 上准备一台高速服务器,或其他 Internet 高速机器来解决此问题。假如已经部署了一台高速服务器,速度为 100Mbit/s,并且上行和下行均可达到此速度,则诊断方案设计如图 7-5 所示。

请先了解图 7-5 的结构,

图中有三个机器:

用户客户端机器(简称 A):普通的客户端。

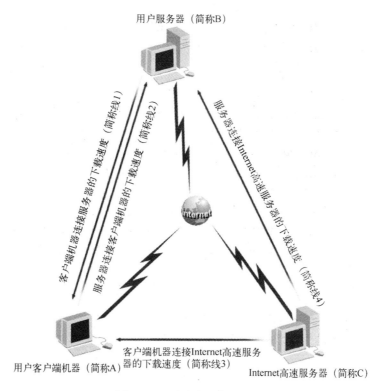

图 7-5 网速诊断方案示意图

用户服务器（简称 B）：安装产品服务器的机器。

Internet 高速服务器（简称 C）：专门提供的一台高速服务器（或其他 Internet 高速机器），服务器不会产生瓶颈，可以测试出连接这个机器的客户端的最大连接速度，和以下四条线路速度：

客户端机器连接服务器的下载速度（简称线 1）；

服务器连接普通客户端机器的下载速度（简称线 2）；

客户端机器连接 Internet 高速服务器的下载速度（简称线 3）；

服务器连接 Internet 高速服务器的下载速度（简称线 4）。

如果客户机连接服务器的速度（即"线 1"的速度）为 50KB/s，准确来说这个速度是指客户端连接服务器的实际速度，网速比较差。通过此值只能确定是客户端或服务端有瓶颈。

基于此局限性，使用我们提供的 Internet 高速服务器来协助客户实现网络速度检测，找出网络带宽瓶颈。具体过程是：

（1）检测出客户端机器和服务器的对连（双向下载）速度。

（2）检测出客户端机器和 Internet 高速服务器的下载速度。

（3）检测出服务器和 Internet 高速服务器的下载速度。

（4）将这四组数据进行两两对比，确定瓶颈所在。

下面举例说明一下整个检测过程。

在本次检测中，第一步的检测结果为：客户端连接服务器的速度为 50KB/s；服务器连接客户端的速度为 80KB/s。

第二步的检测结果为：客户端连接 Internet 高速服务器的速度为 200KB/s。
第三步的检测结果为：服务器连接 Internet 高速服务器的速度为 80KB/s。
检测的数据，如表 7-2 所示。

表 7-2　网络故障检测数据

终端类型	目标类型	实际速度/（KB/s）	结　　论
客户端机器	Internet 高速服务器	200	客户端最大下行速度为 200KB/s
服务器	Internet 高速服务器	80	服务器最大下行速度为 80KB/s
客户端机器	服务器	50	实际值为 50KB/s，小于与 Internet 高速服务器连接的速度（200KB/s），说明受到服务器速度的限制。同时测试出服务器的上行速度最大为 50KB/s
服务器	客户端机器	80	服务器连接 Internet 高速服务器和客户端机器的速度都是 80KB/s，说明客户端不存在瓶颈。反之，如果此值小于 80KB/s，则客户端可能存在瓶颈（指小于与 Internet 高速服务器连接速度）。 同时检测出客户端的上行速度最大为 80KB/s

以上是整个检测过程。总结一下，以上过程其实就是在外网先准备一个带宽足够大的网络服务器为参照物，对线 1、线 2、线 3、线 4 进行两两比较，确认瓶颈在哪台机器的上行或下行。此方案可以灵活自如地应用于多种实际客户场景。此方案能够非常实用地定位到网络瓶颈产生在哪端，进而针对有带宽瓶颈的机器进行故障解决。